D0930714

user *un*friendly

user
*un*friendly

Consumer Struggles with
Personal Technologies, from Clocks and
Sewing Machines to Cars and Computers

JOSEPH J. CORN

The Johns Hopkins University Press
Baltimore

The Johns Hopkins University Press
2715 North Charles Street
Baltimore, Maryland 21218-4363
www.press.jhu.edu

Library of Congress Cataloging-in-Publication Data
Corn, Joseph J.
   User unfriendly : consumer struggles with personal technologies, from clocks and sewing machines to cars and computers / Joseph J. Corn.
      p. cm.
   Includes bibliographical references and index.
   ISBN-13: 978-1-4214-0192-8 (hardcover : alk. paper)
   ISBN-10: 1-4214-0192-4 (hardcover : alk. paper)
   1. Technological innovations—United States. 2. Human-machine systems—Social aspects. 3. Human-computer interaction. I. Title.
   T173.8.C685 2011
   303.48'30973—dc22   2010051494

A catalog record for this book is available from the British Library.

Special discounts are available for bulk purchases of this book. For more information, please contact Special Sales at 410-516-6936 or specialsales@press.jhu.edu.

The Johns Hopkins University Press uses environmentally friendly book materials, including recycled text paper that is composed of at least 30 percent post-consumer waste, whenever possible.

contents

Illustrations follow page 119

user *un*friendly

Our Marvelous and Maddening Machines

I T IS HARD TO IMAGINE LIFE WITHOUT MACHINES. We awake to the buzz or serenade of alarm clocks and radios, move to the bathroom to prepare for the day, cleansing and grooming our bodies with electric toothbrushes, shavers, and hair dryers. Then it's into the kitchen, where we manipulate more buttons, switches, knobs, levers, and controls to run our coffeemakers, microwave ovens, dishwashers, refrigerators, and other appliances, grabbing breakfast before work or school. By that time, we usually have also turned on a TV set, logged on to a computer, or used the telephone. If we work from a home office we may continue to use these devices, and others such as scanners, printers, and faxes, throughout the day. Leaving the house or apartment, we seldom do so without carrying at least a mobile device such as a cell or smartphone, if not also a laptop or tablet computer such as the Apple iPad. And wherever our destination, we usually get there in an automobile, the most expensive machine in our stable of modern gadgets.

Even when we stay home to do housework or just to relax, we also interact with machines. Rarely is any household chore or yard work done these days without powering up a dishwasher, vacuum cleaner, lawn mower, hedge clipper, or some other technology. In recent years, technologies practically unknown a generation ago, such as snow blowers, power washers, and gas grills, have become standard in many suburban households, increasing domestic equipment inventories. When it is time for relaxation or recreational exercise, Americans young and old climb onto electric-powered stepping machines or treadmills; switch on MP3 players and stereo systems; or watch movies, sitcoms, or sports on television, often aided by a DVD player, or DVR. For increasing numbers of us, however, such more passive entertainments no longer suffice, and we turn on dedicated gaming consoles to participate in simulated sports competition, combat, or espionage. Even if we pursue old-fashioned hobbies or activities such as sewing, woodworking, or reading, we are likely to rely on the latest technologies. We recreate the look of classic hand-stitched quilts, not by emulating the techniques of our great-grandmothers but by turning to computer-guided sewing machines. If we enjoy woodworking, we may praise the values of hand craftsmanship but are also likely to deploy an arsenal of machinery to shape wood in our basement workshops. Even those of us who prefer the simple leisure activity of reading are increasingly adopting those twenty-first-century replacements for physical books, machines such as the Amazon Kindle, Sony Reader Digital Book, or the iPad, to download e-books from the Internet.

All these machines, from clock radios to mobile phones, electric toothbrushes to computerized sewing machines, are "personal technologies," not simply because their use is literally personal, as with a toothbrush, or small enough to be carried or worn by an individual, like a cell phone. Rather, they are personal in a more significant way: people must go out and buy one themselves if they desire the benefits these technologies make possible, and then they must learn to operate the machine and attend to its maintenance, repair, and eventual replacement. If they want to wake up to music, power-brush their teeth, or watch baseball without leaving home; if they want the convenience a car makes possible by being able to leave for a trip at any time, they must first become technology consumers. They must shop for and then purchase a clock radio, an electric toothbrush, a television set, and an automobile.

Personal technologies have always been seductive. They augment our strength and extend our capabilities; indeed, some technologies seem almost miraculous in this regard, giving us previously undreamed of potentials, such as the capacity to talk to people who are far away, to view photographs or films of events that happened long in the past, or to participate in virtual environments. Beyond their utility and convenience, technologies have also provided us with much enjoyment and entertainment. Finally, the mere ownership of some machines, such as watches and automobiles or devices that are new and dazzling, confers status and further adds to their desirability.

Yet while consumers have always found personal technologies alluring, they have long been vexed and confused by such machines. When a radically new device comes on the market, people invariably understand little about it: they don't know how the technology works or how to operate it; even deciding whether they want to buy the thing can be difficult. If they do decide to go ahead and acquire one, shopping for and choosing a particular machine from the many models, types, and brands is yet another challenge. Once consumers select a new machine and take possession, however, they are likely to experience additional pain and struggle. "Some assembly may be required," as the euphemism puts it, before they can use their device; or they may need to complete a complicated setup procedure to configure it. Then to use the machine they may also have to learn new skills and acquire difficult technical knowledge. And once they finally get their machine up and running and have mastered its control, they will, as noted, have to attend to its ongoing needs, cope with its troubles, and arrange for its maintenance, repair, and eventual replacement. To be sure, consumers do not struggle similarly with every new technology: in the past they have assimilated some machines with little or no difficulty, while others have tyrannized them for decades. And as we will see, early adopters invariably suffer more than later consumers who buy improved and refined versions of a technology. Yet it is the nature of machines that things fail or go wrong, and the more complex the technology the more possibilities there are for consumers to struggle and experience frustration.

A 1996 study on the subject of "technology and the revenge of unintended consequences" spoke of how things can "bite back" and take revenge, so to speak, on their human creators. A typical example came from medicine, where the introduction of a chemotherapy treatment proved effective for some cancers but induced other, equally malignant tumors. Such a "revenge effect," as author Edward Tenner termed it, was wholly surprising and obviously unintended, but the struggles of untrained consumers with complex personal technologies, while surely not intended by manufacturers, should not come as a surprise. Such struggles are, in fact, understandable and even predictable, an inevitable consequence of modern life.[1]

The "new" technology that has recently been the most maddening—and simultaneously the most marvelous—is the personal computer, which first appeared on the market in the late 1970s and early 1980s. As one who got his first personal computer in 1982, I well remember the painful struggles involved in learning the complex and esoteric protocols and procedures connected with the use of the technology. Little in previous experience had prepared me or most others for dealing with this mysterious and complicated new machine. Computers were fabulously empowering and could do many wonderful things, but as we will see, consumers went through hell assimilating them into their daily routines and becoming comfortable with them.

IT WAS A PROBLEM I HAD WITH A NEW COMPUTER PRINTER that ultimately became the inspiration for this book. At the end of 1986 and on the eve of spending sabbaticals in Washington, D.C., my wife and I had bought a new HP LaserJet 4M printer using our faculty discount at Stanford University, in California, where we worked. Our old Epson printer still functioned, the one we had got along with our first computer four years earlier, but it could not support both my IBM and the new Macintosh computer that my wife had recently bought. The HP LaserJet, alternatively, could be networked with the two disparate platforms. Just before Christmas we pulled up at our rental apartment in Washington, having driven cross-country in a U-Haul truck loaded with books, papers, luggage, and the new printer still unopened in its factory carton. We battled the heavy box up two flights of stairs, and I excitedly unpacked the machine, read its instructions, and started the setup process.

I inventoried the cables, manuals, paper tray, ink cartridge, diskettes, and other items that came in the box with the printer. Placing it on a table, I installed the print cartridge, loaded the HP software onto both computers, and cabled everything together. I then turned on the power, sat down at my IBM, typed a few words, and tried to print my first document. Nothing happened. After trying unsuccessfully a few more times, I tried printing from my wife's Macintosh. Again, zero results—well, not exactly, for the Mac screen did display the unhelpful and obvious message "Printer not responding." After repeating the same steps a few more times, I tried turning the printer off and then back on and then did the same with the two computers. Still the printer still did not print. I checked the various cable connections, but they were tight. I also reread the start-up instructions and fiddled with the buttons on the printer's control panel.

Frustrated, I decided I needed help but was not sure who to turn to. The capable IT staff at Stanford were too far away to consult, so I visited a local computer store, where a seemingly knowledgeable salesperson assured me my problem was that I had the wrong cable—either a serial when I needed a parallel cable or vice versa, I've forgotten—and willingly sold me one. He was wrong; the new cable didn't resolve my trouble. I then talked by phone with somebody in Hewlett-Packard's technical support division. He was very nice but could not tell me what was wrong, though he reassured me that the printer was designed to work in a mixed-platform environment like mine.

Late the next day, after again rereading the relevant sections about connecting the printer to a Macintosh in the HP user's manual, I discovered that I was supposed to change one default setting when using both the printer's serial and parallel ports, but I had not registered that point on my initial (and probably too hasty) readings of the manual. Once I found the problem, correcting it was easy; its settings now adjusted to its liking, the machine worked perfectly, rapidly cranking out pages of gorgeous copy.

My struggle, however much my own fault, paled when compared to the ordeals some technology consumers have endured, yet the experience produced anxiety, frustration, bewilderment, and even anger. And it called attention to the general problem whereby technically untrained people purchase, use, and grow reliant on complicated machines about which they know little. I recall the worries that flitted through my mind as I tried to get the printer to work: Had I bought the right machine? Could the LaserJet 4M *really* be networked with both an IBM and Macintosh? Did I have all of the accessories or attachments needed to do what I wanted to do? And what if the printer were defective? How would I ever know? When I first plugged the LaserJet in and turned it on, it had made noises unlike anything I'd ever heard from our old Epson printer, which was totally silent except when actually printing. The new machine gave a veritable concert of electronic hums and whirrs when first connected to a power source. How was a computer novice and first-time owner expected to recognize the normal voice of the laser printer? I also worried that I might have screwed up the machine while randomly pressing the various buttons on its control panel. As I reflected further on my experience, however, I wondered why the Hewlett-Packard manual—actually, there were *two* of them, a user's manual as well as an equally thick technical tome—did not employ clearer language to take into account the minimal knowledge users like myself tended to bring to computing. Eventually, it struck me that my experience was one that was widely shared. Although the particulars varied individual to individual, virtually everyone who adopted this new and wonderful technology of computers endured similarly intimidating and painful moments.

This was certainly true at Stanford University, where I was teaching in the early 1980s when I got my first personal computer. The school had decided that its humanities faculty were missing out on the pedagogical and research potentials of computers and offered to subsidize any of us who adopted the new technology. Those who took up the university's offer tended to have little or no technical background, yet we were cautiously excited about the prospect of using the machines in our writing and teaching. I recall that my humanities colleagues and I talked about computers quite a lot, considering that we really didn't know very much about them. We pondered the impact computers might have on our scholarly productivity (overestimating it) and wondered whether we ever would give up our pencils and pads, typewriters and carbon paper (here we failed to imagine a typewriter-less future). We also discussed the issue of how the use of these machines might affect writing, both our own and especially that of our students—privileging paragraphs and jumpy short passages over continuous and smoothly flowing narratives, perhaps, or encouraging prolixity. But most of all we shared horror stories and tales of struggle. We complained about complex and illogical software and the bewildering routines required by DOS, the operating system used by the IBM computers that the university helped us buy. We shared tales of woe, like my relatively minor if embarrassing effort to set up a new printer and those more apocalyptic sagas of entire manuscripts lost due to sudden and mysterious computer crashes.

Such experiences suggested to me the idea of a history of the relationship consumers have had with the technologies they purchase for personal use. I was particularly interested in the stories of early adopters, the first generation of users. In 1982 the personal computer had only been in existence for about five years, and so we faculty who acquired the subsidized machines at Stanford, like many others elsewhere, were not only early adopters but most of us were also total novices in regard to computing. Furthermore, the fact the technology itself was in its relative infancy all but guaranteed that we users would face many calamities and considerable confusion. As a historian I suspected that the personal computer was not the first technology to have put adopters through this kind of hell, so I began looking for earlier examples of machines that were also maddeningly challenging but that had also become popular consumer products and to examine the experiences of those who bought and used such technologies.

My experience with computer equipment provided a user's perspective for such an inquiry. Although my introduction to personal computing had been eased by my university's making the difficult technical choices as to what machines and software I should buy, I knew that acquiring new technology was itself often a bewildering and confusing experience, so the experience of shopping and choosing would be part of my inquiry into the history of consumer-technology relationships. I would also look at the challenges of learning the new machines, that is, how to operate them. Because computers often crash or behave unexpectedly, I also would analyze the way consumers become troubleshooters, diagnosing their machines' problems and trying to get them to behave normally again, all of which are also part of our experience with personal technologies. Finally, because when they were confused or in need of help, technology consumers often consulted printed instructions or manuals, I wanted to consider the relationship we have to machines as readers. Yet those verbal and visual representations of a machine-on-the-page, so to speak, could be more confusing than the device itself, or at least this was the case with the user aids Hewlett-Packard included with my laser printer. Taking a close look at the publications manufacturers included with earlier personal technologies would illuminate the consumer experience. Most generally, I hoped to discover which other technologies might have rivaled personal computers in their capacity to infuriate owners, and thereby to provide a broader context for understanding this familiar feature of modern life.

My first forays into libraries made it clear that over the years Americans had found headaches aplenty in their earliest encounters with personally owned machines, starting with clocks and sewing machines in the nineteenth century and extending through automobiles and personal computers in the twentieth. I decided to focus my book on the early adoption of the automobile but also to touch on technologies that came before and afterward. Such a focus would demonstrate the varieties of difficulty and frustration—or lack thereof—that people have experienced in assimilating new machines into their lives and provide a sense of the longer history of struggle entailed in technology consumption. I also decided to consider a few new technologies that were not problematic at all, ones that consumers appear to have assimilated without difficulty, like hand-powered washing machines in the nineteenth century or electric toasters and cake mixers in the twentieth. For various reasons, such technologies seem to have been user-friendly from the outset, or at least no record survives of consumers' struggles or exasperation with the technologies.

I chose to make the automobile the center of this study because no new technology ever sold to ordinary consumers, prior to the appearance of personal computers, was so daunting and difficult to learn. From its arrival in the marketplace around 1900, the automobile, or horseless carriage as it was first called, exemplified the marvelous yet maddening qualities associated with complex personal technologies. People celebrated the convenience, pleasure, social prestige, and even the thrills their automobiles gave them, but at the same time they suffered singular bewilderment and annoyance in coping with the technology's dizzying complexity and finicky foibles. The complexity began with shopping for a machine, for consumers had to choose from a confusing array of technical designs and types of automobiles—touring or runabout, steam or electric, one or multiple cylinder, and so forth. In making a purchase it was hard to know exactly what one needed, for most early cars sold were incomplete in many respects; they lacked windshields, tops, trunks, and headlights, items now long considered essentials but at the time not included as standard equipment. When operators then first got behind the wheel or tiller of their new machines, they confronted a host of additional hurdles in learning to operate and manage their vehicles. Just starting the engine involved complicated procedures, and once the car was running, operators, like machine tenders in factories, had to monitor ever-changing conditions, steering the car, controlling its speed, and avoiding mishaps, and manipulating various controls to keep the machine running. Because early automobiles were easily damaged and prone to all kinds of mechanical troubles, which often occurred far from any garage or help, motorists also learned to recognize and diagnose the machine's basic ailments. Owners would then get out their tools and tinker under the hood, and greasy hands and skinned knuckles became almost unavoidable accompaniments to being an automobilist, whether just to change a tire, clean a spark plug, or tighten a slipping fan belt. The absence of a national network of gas stations, repair shops, and parts supply stores in the early days motivated owners to become more self-reliant while adding to their challenges. Some early adopters were so fascinated by the innards and workings of their automobiles that they became eager students of mechanics, while others reluctantly learned more than they would have liked about such matters; it was the price consumers paid for the benefits of motoring.

Facilitating my study was the vast paper trail early adopters of the automobile left behind. These men and women wrote extensively about their trials and tribulations, creating a rich archive of published and unpublished material. Just as my generation talked and wrote about its struggle with personal computers, so too did automobilists, as they were known, talk and write about theirs. Their writings constitute a treasure trove for historians wishing to peer over their shoulders, so to speak, and observe their difficulties and struggles. Many early motorists also kept diaries or logs detailing their car's operating expenses, the troubles it experienced, and the work that was done on it. Others wrote articles reporting what it was like to drive in the mountains or to own and operate different types of vehicles—steam, electric, or internal combustion—while physicians wrote about the efficiencies that came by replacing an old "hay motor" with a horseless carriage to make house calls. Owners generally related their "war stories," that is, the battles won over recalcitrant and balky engines or heroic roadside repairs that enabled them to limp home and avoid the ignominy of having to be towed by a farmer and his team. Lengthy trips by early motorists frequently prompted the publication of a memoir or account, recording what seemed like every flat tire, broken spring, and other mechanical mishap encountered along the way. Also evidencing motorists' struggles, though not at all personal testimony, are the mountains of technical literature devoted to understanding and coping with car troubles. Automobile manufacturers published technical bulletins, shop manuals, and owner's manuals, while many engineers and self-proclaimed motoring experts wrote how-to articles and books about buying, operating, and maintaining automobiles. Finally, the mass media, including family magazines like the *Saturday Evening Post* and women's periodicals like the *Ladies' Home Journal,* also regularly published articles advising readers how to cope with this new and complex and highly demanding technology. All of this literature attests both to the fragility of early vehicles and their almost infinite capacity to give operators trouble as well as to the struggling car owner's craving for additional information.

For a couple of reasons, then, the early adoption of the automobile offers us a paradigmatic example of what I call technology consumption. Consumers who adopt new technologies have quite different experiences, and encounter different problems, compared to those who purchase and use groceries, furniture, clothing, or other nontechnical goods and products—things that are familiar and relatively unchanging. I call this consumption traditional to distinguish it from technology consumption. Although consumption generally has attracted considerable attention from anthropologists, sociologists, economists, and historians, such scholarship often blurs or altogether ignores differences among the things consumed. All different types of acquisitions—of shoes, cars, books, pianos, lamb chops, patio chairs, cameras, window shades, handbags, and the myriad other goods and products exchanged in the marketplace—get lumped together and treated as a single phenomenon: "consumption."[2] Aggregating goods this way is often essential and valuable, as when studying the economy or looking at poverty or when charting the rising importance of department stores in nineteenth-century retailing. But all consumption is not equal, and treating it as such tends to ignore the particular kinds of experiences consumers have, and the confusion and struggle they sometimes have with machines but rarely with other kinds of goods or products. Another tendency of most scholarship on consumption is to focus exclusively on shopping and buying, as if the use of goods purchased—to say nothing of their maintenance and repair—were unrelated to consumption. Thus scholars have examined the part advertising and store display have played in constructing consumer desires, while others have looked at the experiences, settings, and practices of shopping.[3] These are important and essential aspects of consumption, but by no means do they constitute the whole story. Consumption does not end in the mall parking lot, as historian Susan Strasser quipped in a recent review of scholarship on the subject.[4] There is much work to be done both before and after consumers bring home the bacon, so to speak, though scholars often have ignored such work. "The responsibilities of the consumer are no mean task," observed the great historian of American culture David Potter, some fifty years ago, and "to handle them successfully, a person must be something of a dietitian, a judge of the quality of many goods, a successful planner, a skillful decorator, and a budget manager."[5] Although his comment was skewed toward the era's emphasis on the housewife as a consumer, it reminds us that consumption is often a multifaceted activity.

This is all the more true when the objects of consumption are personal technologies. The experience of consumers here is more complicated than when traditional goods are involved, so let us briefly compare technology consumption with traditional consumption.[6]

It is obvious that most consumption begins with shopping, an activity that is often more difficult than those who deride it as largely impulsive behavior might imagine. Even buying groceries, things that are familiar and traditional, argues sociologist Sharon Zukin, can be complicated and anxiety producing. "In the store you face competing brands of chicken that are all healthy, guaranteed to taste good, and carry the endorsement of cultural authorities," she writes in a recent study. "But at home, you've got a grocery budget, a spouse who likes a leg of lamb, and a daughter who is a vegetarian. Besides worrying about the chicken, you worry that you, as a shopper, will fail the test. You're immobilized by the anxiety of making the wrong choice."[7] When shopping for more personal items, such as "the perfect pair of *leather* pants" Zukin discusses in a later chapter, a woman named Cindy is described as confronting all those anxieties along with added complexities. Cindy discovers there are myriad brands of leather pants, and that they come in many types and styles; the quality of their leather also varies. After considerable looking and trying on many pairs, Cindy finds one that fits, looks good on her, and is not too costly. Her success demands persistence, patience, and physical endurance, and Cindy learns things about leather pants she had not known before. But none of the knowledge she gains, I would argue, whether about pants or leather, was hard or difficult to understand; much of it, in fact, Cindy already had. Zukin explains that "shopping is most intensely a learning process" and "lifelong learning begins in childhood." Little girls go to stores with their mothers and begin to learn the variety of clothing types—pants and tops, skirts and dresses—and along the way they develop some sense of cost and value. Even more so do they form notions regarding fit and style, first taking cues from parents but quickly substituting those gleaned from peers.[8] Long before Cindy went in search of perfect leather pants, then, she knew a great deal about pants generally and had formed expectations as to how they should fit her, what they should look like, and what a particular level of quality and workmanship should cost.[9] The only thing she did not know much about was leather, Zukin tells us, but after visiting a few stores, Cindy became able to distinguish better leather by feel and appearance. In short, although her quest for leather pants was protracted and arduous, she never was bewildered in the way shoppers for new technologies can be.

Technology shopping is almost by definition vexing, as the products are novel and often the result of cutting-edge technical and scientific advance. Whether shopping for automobiles in the early 1900s or computers in the early 1980s, consumers could not draw upon childhood experience for guidance as could Cindy in seeking leather pants. Consumers often found it difficult to grasp the usefulness of such new technologies. What exactly can the machine do and how might it be used? The uses of new technologies are often novel, specific as to context, and must be learned—often by actual use and through trial and error. And unlike the objects of traditional consumption, the functions of technology can change rapidly as devices evolve. Once technology consumers decide which machine to buy and come to own it, they are likely to confront an unprecedented variety of problems that have little parallel in traditional consumption. Buyers of leather or cloth pants might lose a button or tear a seam, and shoppers might not exactly know how to prepare their chicken, but everybody grows up with at least a general awareness of the uses for, and the management and care of, clothes and foodstuffs. And if they lack such knowledge they know where to turn; they can ask a friend or relative, or in the case of food preparation consult a cookbook. And they also can readily describe their problem or need—the seam has ripped and needs sewing, or they need a recipe for cooking chicken. Technology consumers, however, face greater challenges. Learning to use their new machine might demand considerable study and practice, and if the machine fails to work or perform as expected, consumers often neither can describe the problem nor know what to do to rectify it. Because they start with so little understanding or knowledge of their machines, then, technology consumers are often forced to read and study instructions or owner's manuals, texts that are seldom provided or needed for users of traditional goods.

Technology consumption, then, although it begins with shopping as does traditional consumption, imposes additional burdens on consumers as owners and users. Unlike foodstuffs, which are rapidly consumed or eaten, machines endure and must be nourished and maintained, like domesticated animals. And unlike clothing, technologies often require significant learning to be effectively used at all. Every aspect of consumers' relationships to such technologies, then—from shopping for them to using them, and from coping with troubles to seeking enlightenment through reading instructions—tends to be more challenging and difficult than are consumers' interactions with traditional goods. Personal technologies, in short, can be user unfriendly in ways that traditional goods are not.

The relationships we have with personal technologies therefore subject us to a kind of servitude: we must labor on behalf of our machines if we want them, in turn, to be there for us. Advertising hype wants us to believe that every new mechanical or electronic marvel will save us time and labor and even liberate us from worry, beliefs strongly nourished by our culture's faith in improvement and optimism about technology. While the benefits and pleasures of owning and using machines are undeniable, so too are the demands they force on us as consumer-users. If we want the benefits a technology promises, we must learn to do things the machine's way—we must set our printer's defaults correctly, fill the car's tank with gas, procure the correct battery for the garage door opener or bulb for a flashlight. Today's automobiles literally imprison us in that, before we can get out, we often must put the car in park so as to be able to open the door. In such cases, of course, it is the design and construction of a technology that largely determines the basic actions people must perform to control and use the device.[10] Although humans are putatively the masters of their technology, they must do its bidding if they expect it to serve them. In doing so, by attending to the demands and needs of our machines, we humans in effect become their slaves. To be sure, this is a relatively soft and gentle servitude, unscarred by the horrific cruelties of chattel slavery, where humans enslaved their brethren. And this servitude is obviously voluntary. But for the consumers who live every day with machines, it is easy at times to feel more like captives than captors.

TECHNOLOGY CONSUMPTION, or the purchasing of manufactured machines and devices for personal or domestic use, is a recent phenomenon, barely more than two hundred years old. At the time the United States declared independence during the American Revolution, and even as late as 1800, most of the country's population was relatively self-sufficient. People lived on farms, grew crops and raised animals for food, and produced goods such as yarn, clothing, butter, and soap in their own households. At the same time, even on the frontiers, eighteenth-century Americans did some shopping. Since early in the colonial period, they had been purchasing yard goods and hardware, tools and building supplies, and foodstuffs like tea, coffee, wine, and molasses from merchants and peddlers. As early as the 1600s, in fact, historians speak of the "culture of consumption" that entwined Americans and Europeans, causing them to look to the marketplace not merely for luxuries but also for many staples. Yet few people owned or had in their homes anything we would call a machine or technology.

With few exceptions, consumers could not purchase machines in the eighteenth-century marketplace, although we must be more precise as to what we mean by a machine as well as what we mean by technology, as these are concepts I often use interchangeably. "The term 'machine' presents an almost hopeless problem," wrote historian Abbott Payson Usher of its definition in a pioneering study of mechanical inventions published in 1929. Usher believed the frequently made distinction between a machine and a tool, the latter being an implement "for the direct execution of certain kinds of work, notably cutting, striking, or rubbing," was insufficient because many former hand tools had been powered without markedly changing what they did or how they performed their work. An electric cake mixer and gasoline-powered lawn mower are examples of simple tools, a whisk and scythe respectively, that when powered by motors became machines, even though their function—whipping and cutting respectively—remained unchanged. So too was the earlier hand-pushed lawn mower, widely adopted following the U.S. Civil War. Unlike a simple tool such as a scythe, a nonpowered lawn mower had a "transmitting mechanism," as Karl Marx called the means by which machines harnessed or transmitted muscle power to a tool to perform work.[11] In the following pages, therefore, we will use the term *machine* to refer to manufactured products that either have autonomous power sources—springs, electric motors, or other kinds of engines—and/or utilize some mechanism to transmit power, whether from wind, muscle, steam, or any other source, to perform tasks desired by their consumer/users.

The meaning of the term *technology* is perhaps even more elusive and difficult to pin down. In the 1990s the word increasingly became associated with computers and devices related to information processing. Earlier, the word had denoted a much broader range of things, techniques, and practices. Anthropologists and other scholars, including historians of technology like myself, still use it to refer to the many varied artifacts humans have made to augment their strength, survive against nature, or extend their capabilities—from simple tools like spoons and hammers to machines like lawn mowers and clocks to constructions such as skyscrapers and concert halls—all are technologies in this broader meaning of the word. It thus becomes obvious that Americans before 1800 owned and used technologies: they had wagons for hauling, rope and tackle for lifting, and houses and barns to enable them and their animals to survive the New World's harsh winters.[12] To make the interiors of their homes more habitable in cold weather, they improved on older fireplace technologies. One who did so was Benjamin Franklin, who invented the more efficient stove that carries his name. As an inventive octogenarian, he benefited from another technology common in the eighteenth century: eyeglasses, an invention of the Renaissance that enabled consumers to see and remain productive. Also evident in eighteenth-century homes were communications technologies in the form of books, along with paper and pens, the last including old-fashioned quill pens and the recently invented steel-nib pointed versions. People also owned and used entertainment technologies: the well-to-do had pianofortes and harpsichords, while common folk and even slaves possessed fiddles and banjos. Every colonial household also had technologies for preparing, storing, and serving food, including plates, trenchers, mugs, jars, iron fireplace cranes, forks, and knives. Heads of households traditionally owned those technologies related to their trade or occupation, the most common being farming. Hay rakes, pitchforks, shovels, hoes, axes, adzes, harnesses, and yokes for animals were just a few of those owned by most farmers. Many in the colonial and early national period also kept a firearm, although the extent of handgun, musket, and rifle ownership—long assumed to have been nearly universal in a pioneering and hunting society—has recently been challenged with the claim that "gun ownership was exceptional in the eighteenth and early nineteenth centuries, even on the frontier" and that the widespread adoption of firearms came only after industrialization.[13]

The most obvious machine of eighteenth-century America was the mechanical clock. Developed in Europe during the fourteenth century, clocks exemplified technical ingenuity and mechanical craftsmanship. Because clocks were intricate and handmade, however, they were expensive, and before 1800 they would be found only in the homes of relatively affluent families. The spinning wheel was not thought of as a machine at the time, and might not strike us today as much of one, yet the devices were widely diffused in eighteenth-century American homes. One study of the Connecticut River Valley in 1750 showed that over 80 percent of the households in the area owned one.[14] A relatively simple machine, powered by hand and constructed largely of wood, a spinning wheel was used by women to transform raw cotton, flax, or wool fibers into yarn. This might then be woven, usually by another family member, into homespun or sold to merchants who contracted with factories to loom it into cloth for the market.

Of course, in eighteenth-century America nobody referred to a spinning wheel, clock, gun, rake, trencher, or any other artifact as a technology. Few people had even heard of the word, and indeed its meaning then was quite different from the one it would later acquire. At that time technology meant the knowledge of a particular trade and the theories, practices, and recipes employed therein. One might speak of the technology of glassmaking, say, as a shorthand way of summing up everything known by a skilled glassmaker. In the latter half of the nineteenth century, *technology* took on a couple of new meanings. The first was the broader anthropological definition already alluded to, as when people refer to the entire gamut of human-created tools and other useful artifacts as "technologies" or "human technology." The second usage confined the term to a smaller subset of material objects, mostly machines and large structures like bridges, dams, or skyscrapers, things commonly resulting from the activity of engineers. Additionally, this use of the word subsumed the knowledge required to design and build such things. It is this last meaning that underlay the name given in 1861 to a new educational institution to train engineers, the Massachusetts Institute of Technology, or MIT.[15] Both of these nineteenth-century meanings of *technology* reign today, although recent usage increasingly has inflected the word with the third meaning, already noted, narrowing it further to the world of digital and computer-related devices.

The most important point to make about pre-1800 tools, machines, and artifacts in America has nothing to do with what people called them or with the terminology they understood or recognized. Rather, it is the fact people did not experience technologies as bewildering or exasperating. This is because all of those pre-1800 artifacts—the spinning wheels and hoes, cooking utensils and violins, paper, pens, and guns—all had been around for as long as people could remember, often for many generations, if not centuries. These things were familiar. People's parents and grandparents owned them, if not the actual gun or harness or barn, then one just like it, for artifacts were relatively unchanging, year to year and even century to century. A brand-new spinning wheel, hay rake, ceramic pot, ax, musket, or pair of eyeglasses in 1800 differed little from one made in 1750 or even 1700. Preindustrial artisans did not rapidly innovate or change their practices, so whatever changes in design and construction had occurred over the years would have been minimal. Similarly, the uses and ways of using technologies resisted change. Daughters learned from mothers how to use spinning wheels and cooking utensils, while sons learned from fathers to handle rake, ax, and firearm; the techniques and skill requirements for using such things, and the knowledge related to use, were largely static. And when a traditional technology failed—when a tool handle split, a crock broke, or a roof leaked—the means of repairing or replacing the object were also well known and understood. Owner-users might not themselves know how to make the repair, and often they were incapable of fabricating a replacement. But they knew whom to turn to in such cases. In short, before 1800 people lived in a world where material novelty was largely restricted to art or fashion. It was also a world where people were unchallenged by their technology. They were not forced to learn new things or acquire new habits, because even "new" technologies, at least in their design, construction, and use, were also old and familiar. Material life in 1800 was rich and might involve much consumption, but it was also traditional. It did not require the incessant and often challenging adaptation to novelty characteristic of more recent technology-based living.

These traditional patterns were upended, however, during the industrial revolution of the nineteenth century. Driven by the development of steam and internal combustion–powered machinery and the introduction of assembly lines and mass production, and facilitated by faster forms of communication and transportation along with the rise of corporate forms of business, industrialization rapidly transformed material life in the United States. Consumers initially benefited from industry's ability to produce more plentiful and less expensive traditional goods, such as textiles and nails, hand tools and shoes, foodstuffs and building supplies. More relevant to our story, however, was that industrialization for the first time enabled masses of consumers to buy factory-made machines: clocks and watches, sewing machines and bicycles, meat grinders and lawn mowers, washing machines and phonographs, and many other innovative personal technologies. As people shopped for, used, and lived with these newly available machines, they experienced the confusions and bewilderments—along with the pleasures and satisfactions—we today associate with personal technologies. Nineteenth-century Americans came to understand the frustrations of choosing which machine to buy, learning to use it, and puzzling over incomprehensible instructions as they sought to diagnose its maladies if it didn't run. Becoming accustomed to using machines and often dependent on them, consumers soon discovered the modern phenomenon of technological obsolescence. Devices bought but a few years earlier might still work but had become less desirable than and often decidedly inferior to the latest versions of the technology. Or old machines that wanted but a single replacement part could no longer be repaired because the manufacturer no longer produced that item, thereby forcing consumers into costly upgrades. The age of personal technologies had arrived!

IN THE FOLLOWING CHAPTERS we trace this history of people's interaction with their personal technologies, starting with the emergence of technology consumption in the nineteenth-century United States, which is the subject of chapter 1, "The Advent of Technology Consumption." That story begins around 1800 with the advent of factory-made, inexpensive clocks that made mechanized timekeeping affordable for all classes of Americans, not just the affluent who long had owned handmade watches or tall case clocks. Manufactured mostly in Connecticut and peddled across the country, these inexpensive timekeepers for most people were the first machines to enter their homes, the first personal technologies, as we are using the term. Clocks were followed at midcentury by sewing machines and a variety of other appliances that fastened Americans into their new relationship with technology. To be consumers now meant being operators and tinkerers, adjusting and fiddling with devices; living with machines also meant becoming readers, puzzling over manufacturer's directions and owner's manuals. These consumers' pioneering interactions with industrially produced technologies previewed the ways we today relate to our computing and communications gadgets.

The four chapters that follow the examination of the nineteenth-century origins of technology consumption detail the consumer's experience with early automobiles, the most difficult and vexing machines sold to ordinary people for personal use prior to computers. In chapter 2, "Buying an Automobile," I treat the early car-shopping experience; in chapter 3, "Running a Car," I look at automobile owners as operators, analyzing the struggles novices went through acquiring the necessary knowledge and skills to run their new machines. In chapter 4, "Tools, Tinkering, and Trouble," I consider car buyers as tinkerers, wielding tools and confronting the many troubles that arose under the hoods of their machines. In chapter 5, "Reading the Owner's Manual," I wrap up the study of the automobile by discussing a most archetypal aspect of technology consumption: learning by reading, the experience of consulting instructional texts published by the manufacturer or others. Often it is only as a last resort, after everything else has failed, that frustrated technology users turn to the book in quest of answers, hoping the machine-on-the-page somehow will be more comprehensible than the actual three-dimensional one in front of them. Automobiles were not the first machines sold with instructions, but their complexity rendered such texts mandatory and they were the first technology to spawn a vast web of independently published books and articles offering information-starved users help and advice.

Finally, after the chapters on the automobile consumer—as shopper, user-operator, tinkerer-diagnostician, and reader—in chapter 6 I look at the early adoption and use of the personal computer, the consumer technology that got me thinking about technology consumption in the first place. I consider this most influential personal technology against the background of consumer experience with other post–World War Two electronic and electrical devices such as television and other household appliances. In the book's epilogue, "The Technology Treadmill," I look at how digital computing technology has spread into ever more machines and devices, continuing to stymie consumers and ensure that *user unfriendly* remains in our vocabulary.

Chapter 1

The Advent of Technology Consumption

TODAY WE TAKE FOR GRANTED THE MANY TECHNOLOGIES that keep track of time: not only our clocks and watches but thermostats, microwave ovens, alarm systems, personal computers, and mobile phones. Timekeepers are everywhere, but once we learn as children to tell time these machines become part of the background of our lives, and as users we do not think of them as challenging. When inexpensive mechanical clocks first arrived in the early 1800s, however, they could seem baffling and complicated. It was not that clocks were a new invention; mechanical timekeepers had appeared independently in both Europe and China in the thirteenth century, and the colonizers of North America had brought clocks along with the knowledge of how to make them at the beginning of the sixteenth century. But at the start of the nineteenth century, few Americans owned such devices or the miniaturized versions called watches. Timekeepers were still costly, being by nature intricate devices that had to be painstakingly constructed by skilled craftsmen. A 1799 survey in Connecticut, a state heavy with resident clockmakers, revealed that only one out of every twelve households possessed a clock, usually a tall-case version designed to accommodate the long pendulums that ensured accuracy.[1]

Colonial Americans may not have owned many clocks or watches, but they highly valued the commodity we call time and possessed a keen awareness of its value. Puritans in New England, following more general Protestant thinking, believed time was God-given. Human beings were divinely granted a moment on earth and it was incumbent on them to not waste it, to manage it well.[2] Not surprisingly, Puritan preachers timed their own sermons using an hourglass that they placed on the pulpit so it was visible to the congregation. If they did not finish their remarks before the sand in the glass had run out, preachers would tell their audience they were going to "take another glass." Then they would simply turn the hourglass over and continue to talk, ignoring the irony inherent in timing themselves while making up their own rules about time. Clearly, neither Puritan preachers nor colonial congregations had yet become slaves to clock-driven punctuality; that would come later in the nineteenth century, facilitated by the mass ownership of mechanical clocks and watches and by the railroads, which regulated their train schedules and timetables on precisely kept clock time.[3]

Most colonial Americans approximated the time of day from the sun, by reading the shadows cast by trees or buildings on the ground; some built outdoor sundials for this purpose. By observing when the sun was highest in the sky, people also knew when to break for lunch, noon being the most important moment between daybreak and dusk. Those who lived in towns and villages had access to mechanical time in the form of public clocks, mounted on the steeples of churches or the towers of town halls, mills, and factories. They could read the large faces of such clocks at a considerable distance, and from an even greater remove they could hear a clock's bells, which chimed hourly and usually every half hour or even fifteen minutes. People regulated their lives by these clocks, heading to church for worship, home from the fields for the noon meal, or to the mill or factory at the beginning of the morning shift. Given the strong cultural imperative not to waste time and the widespread understanding of clock time in preindustrial America, it is understandable that when the price of mechanical timekeepers fell, consumers purchased the devices for their own homes.

As part of the wider industrial revolution affecting manufacturing and transportation between 1800 and 1850, both the design of clocks and their means of fabrication changed dramatically. Clocks became affordable for almost everybody. Inventor-entrepreneurs such as Eli Terry of Plymouth, Connecticut, transformed the production and assembly of clock components from a skilled handcraft into an industrial process. In his factory, Terry harnessed water power to mechanize the making of gears, pinions, and other clock parts, at first using wood rather than the traditional brass. The machines he devised could be operated by workers with limited skill and training. Terry also invented various gauges and measuring devices that allowed a relatively unskilled laborer to produce parts to exacting tolerances so they could be easily assembled without time-consuming fitting or filing. According to industrial historian David Hounshell, Terry contributed to the new methods of production that envious Europeans would soon be calling the "American System" of manufactures, reliant on innovations in materials and product design as well as special-purpose machines rather than worker skill.[4] Terry's wooden-movement clocks, along with those made by his competitors, were relatively crude and inaccurate. But they were inexpensive and found a ready market in a country increasingly in a hurry.

Most of those who bought clocks in the first half of the nineteenth century probably did not shop for them in the sense of looking at different models and making a choice. The Yankee peddlers from whom they purchased tended not to carry much inventory in their wagons or on their backs, but those itinerant salesmen served scattered rural populations with limited access to retail establishments. Carrying clocks up and down the eastern seaboard and across the Appalachians into the valleys of the Ohio, Tennessee, and Mississippi Rivers, peddlers constituted the major channel for distributing this first household technology.[5] "Peddling clocks became an industry in itself," writes historian Joseph Rainer, and by the 1830s peddlers had "saturated" even the plantation South with inexpensive timekeepers. "Wherever we have been in Kentucky, Indiana, in Illinois, in Missouri, and here in every dell of Arkansas and in cabins where there was not a chair to sit on," noted a nineteenth-century English traveler, "there was sure to be a Connecticut clock."[6]

By the 1840s peddlers were selling new and improved models. New England clock makers had abandoned the use of wooden movements, which, with changing humidity and temperatures, contracted and expanded and finally cracked, causing the clocks to keep poor time or quit altogether. Now clock makers stamped gears from brass sheet and turned pinions from brass rods; increasingly, they substituted springs for falling weights to power clock movements, with the result that new models were smaller, more accurate, and often cheaper. These clocks also represented the first "technological upgrade," as Rainer put it, a phenomenon that technology consumers would become increasingly familiar with in the future.[7] By the time of the Civil War, American households everywhere were likely to own a clock, whether a hand-built heirloom, a still functioning mass-produced model with wooden gears, or one of the upgraded versions with a brass movement. By the final decades of the century, cheap clocks could even be seen on the walls in the shacks of impoverished black sharecroppers in Alabama's cotton belt. Observing this phenomenon, the black reformer and educator Booker T. Washington believed the timekeepers were a "showy" extravagance, as "in nine cases out of ten there would have been no one in the family who could have told the time of day."[8]

Whether or not those who bought clocks could tell time, as first-time owners they also faced the challenges of operating their mechanical timekeepers. Today it might seem silly to speak of operating a clock, for not only do many modern clocks run themselves, powered by batteries or electricity, but the work of winding or setting spring or weight-driven models has for most people long been familiar and second nature. But most purchasers in the early 1800s probably had never seen such a machine up close, let alone handled one, and thus even simple weight-driven wall and shelf clocks could have seemed complicated and mysterious. We do not know this from clock buyers, for people either did not write about the challenge of handling these new machines or their writings have not survived. But we can infer buyers' bewilderment and confusion from studying the clocks themselves, examples of which are in the collections of the Smithsonian's Museum of American History, along with the instructions manufacturers provided with them.

Instructions were usually printed on labels two to four inches on a side and either affixed to the inside or the back of a clock's wooden case.[9] The including of instructions with a manufactured product makes clear that consumers needed help in using the new machines. Typical of these instruction labels is the one attached to an Ives and Lewis wooden-movement shelf clock, manufactured sometime in the 1830s or early 1840s. It is what was called a "thirty-hour" clock, capable of running without winding for a day plus a few hours. The directions bundle the work of winding under the heading "To Keep the Clock in Order," or, as we might say, to keep it accurate and on time. Preceding the directions for winding, however, are those "To Set This Clock Running."[10] I quote that section in full:

DIRECTIONS

TO SET THIS CLOCK RUNNING

Make the Clock fast where it is to stand, in a perpendicular position—take off the hands and face. The face can be taken off by drawing two pins each side the face, which are parallel with two notches on the edge of the face, and then pulling forward.

Oil the points of the brass or crown wheel on which the verge operates. One drop of good oil is sufficient, if rightly applied.

The Pendulum should be put on with the hook from you. Care should be taken not to suffer the key to be put on and turned until the cord is on the pully [*sic*] each side the Clock, and the weights put on. It is most convenient to put the Clock in motion before the face is put on. If it fails to beat equal, it will be necessary to carefully bend the wire which carries the pendulum till it beats equal. If the clock runs too slow, shorten the pendulum; if too fast, lengthen it by means of the screw and slide which runs through the ball.

These directions are filled with technical jargon. In them the manufacturer employed specialized words and expressions that most consumers would not have known. Buyers only had these instructions and the clock for guidance: there is no glossary that defines terms such as *verge, crown wheel, slide,* or even *pendulum,* all of which were most likely foreign to first-time clock owners. They might correctly intuit that "the brass or crown wheel" referred to the single brass wheel in the clock's mechanism—rather than the other wheels and gears, all of which were wood—and properly apply the drop of oil to it. But people who had never before owned a clock or looked into its innards could easily have been confused. Similarly, while it might to us seem obvious that we would face the clock when attaching its pendulum, in which case the phrase in the directions to put on the pendulum "hook from you" makes some sense, to one inexperienced with mechanisms this and similar spatial references in the printed instructions would have been baffling.

Moreover, consumers in the early nineteenth century had virtually no experience following printed directions, for the practice of writing out brief, step-by-step sequences for doing or making something was itself only evolving. The anonymous authors who wrote the clock directions themselves had a lot to learn about what made for a clear instructional text. The sequence laid out in the first sentence of the directions, directing readers to "take off the hands and face" of their newly purchased clock, must have been puzzling. Why, one might wonder, would the factory attach the face and hands only to require the buyer later to undo that work? Only if one read to nearly the end of the first section would the answer become clear: to oil the crown wheel, the clock's hands and face had to be removed. Assuming a reader dutifully followed the directions to this point, he or she might decide it was time to reattach the face and hands. This would have been a mistake, however, as the manufacturer urged users to "put the Clock in motion before the face is put on" so as to make easier the adjustment of the pendulum wire, should that be required to ensure regular beating. Whoever wrote these directions had not yet learned to forewarn readers that there were specific steps involved in setting up the machine and then to number them so users would be able to perform them in proper sequence.

Yet another challenge to this first generation of clock owners was embedded in the seemingly straightforward admonition to apply one drop of "good" oil to the verge and crown wheel of their machines. Folk sayings such as "it is the squeaky wheel that gets the grease" make clear that the concept of lubrication was long familiar, but traditional grease produced from animal fat was too heavy for use on a clock, and oil was not yet a common commodity. In antebellum America, consumers could not just go to a hardware store and buy a small can of oil as we do. Even in the 1830s and 1840s, petroleum-based oil still lay in the future, so clock owners trying to follow instructions could turn to two other possible oils. Vegetable-based oils such as olive, cottonseed, or flax oil were well known, but these were gummy and unsuitable for machinery. That left one other alternative: whale oil. It would have been a fine, light, and truly "good" oil in the sense called for in the directions, prized for illumination as it burned brightly with little smoke. But whale oil was expensive and not easily available in many of the thinly settled regions plied by peddlers selling Yankee clocks. And if available, it was not sold by the drop. So early clock buyers faced a dilemma: Must they go out of their way and spend precious funds to satisfy their machine? Might they make do with a substitute, say a dab of butter or lard? Or could they get away with ignoring the directions altogether? As this example reveals, clock buyers were the first to confront the sort of quandaries over interacting with their machines that have puzzled new technology consumers for over two centuries.

The tasks for keeping their clock "in order," as the second section of the Ives and Lewis directions termed it, were as follows:

> To wind up the weights, put on the key with the handle downwards, and turn towards the six o'clock figure, and keep steady turning till the weight is up, then ease back and take off the key.
>
> If the hands want moving, do it by means of the longest, turning at any time forwards, but never backwards when the Clock is within 15 minutes of striking, nor farther than to carry the longest hand up to the figure XII.
>
> If the Clock should strike wrong in consequence of running down, or other accident, it may be made to strike until it comes to the right hour, by means of pressing down a wire at the left hand of the Clock, directly behind the face, which is attached to the lowest cross piece near the centre of the edge of the plate; repeat the operation, if necessary, till it strikes the right hour.

The key mentioned in the directions refers to the removable tool for winding the clock. A clock key was not exactly an alien device, for it was but a variant of the key used to open a lock, a security technology that went back thousands of years. Nor was the technique for using a clock key that different from that for using a key to work a lock. But in the early 1800s most Americans had little experience with locks and keys. Until industry began to mass-produce them following the Civil War, only the well-to-do were likely to have homes or possessions they kept locked.[11] The humble key (or as here, a device more like a crank) that came with mass-produced clocks therefore initiated many consumers into the practice of wielding a tool to interact with a machine.

Because winding a clock was not at all self-evident, the directions anticipate that the process might be new and begin by explaining how to insert the key or crank into the clock's face. This should be done "with the handle downwards," the text instructs, after which the user should turn the key "towards the six o'clock figure," a locution that suggests the now taken-for-granted notions of clockwise and counterclockwise. Having attempted to avert two possible mistakes on the user's part—those of inserting the key incorrectly and of turning it in the wrong direction—

the directions call for the user to keep turning the key until "the weight is up"—that is, lifted as high as it would go inside the clock's case (the origin of our phrase, to "wind *up*" a clock); this position could be visually confirmed by looking through the front glass of the Ives and Lewis clock and many other weight-driven timekeepers. With clocks powered by coil springs, however, users found it more difficult to know when the clock was fully wound, as there was no visual indicator of that state. Consumers had to feel the spring's tightness in their fingers as they turned the key and learn through experience when to stop. This was tricky: wind the spring too tight and it might break; not tight enough and the clock would run down and lose accuracy before its next winding. The instructions for spring-driven clocks never described just how tight "tight" meant; this was tacit knowledge, something people had to learn by doing as they assimilated the new technology.

Early clock consumers faced another problem that the directions did not address. How could they get their clocks "in order" and set them when no other clock was available from which to tell the time? One way was to listen for the chiming of a public clock or the whistle of a railroad locomotive running on a known timetable and synchronize the clock's time to those distant signals. "Their whistles can be heard so far," Henry Thoreau wrote of the trains running near his Walden Pond cabin outside of Concord, Massachusetts, in the late 1840s, "that the farmers set their clocks by them."[12] In many rural areas, however, clock buyers had to set their new clocks using sun time, either reading it from a sundial or moving the hands of their clock to twelve when the sun appeared to reach its zenith. Some buyers received with their clocks a supplementary set of instructions called an "equation of time." This explained how to convert solar, or apparent, time to mean, or clock, time. Using such equations, however, was challenging, for not only did one need to know the date and month as well as the latitude where the clock was being used, but one had to then work through fairly complex mathematics to solve the equation.[13]

Regardless of how people set their clocks, buyers of this first mass-produced consumer technology not only got a machine that could tell them the approximate time; they also acquired a tutor on matters mechanical. Setting up a clock, adjusting its pendulum, studying the manufacturer's directions that accompanied the device, and using keys to wind up the weights—these tasks were invariably instructive, even if intimidating or puzzling. Clock buyers acquired some of the new vocabulary associated with industrial machinery along with familiarity with important machine components such as gears, bearings, and pinions. The experience gained from winding up clocks proved easily transferable to other devices reliant on clockwork technology that appeared later in the century, such as music boxes, wind-up toys, and phonographs. Because clocks required minimal adjustment and rarely needed repairs, owner-users received little tutelage in tinkering or diagnosing mechanical problems. For that very reason, however, these inexpensive timekeepers inculcated one final lesson. Clocks spread the technical ideal of machines that run themselves with minimal control or intervention by a human operator. People honored this ideal with the maxim "It runs like clockwork" when praising a well-performing and smoothly operating machine or organization. Indeed, whatever struggle buyers experienced setting up their first clock and setting it in motion would quickly be forgotten once it was installed in the parlor or dining room. The machinelike nature of this first personal technology, emitting a regular and soothing "tick-tock, tick-tock," faded as the clock came to seem a natural part of the home, symbolizing familial harmony and social respectability.

OTHER NEW TECHNOLOGIES soon challenged Victorian consumers, most notably the sewing machine. Newly invented and unrivaled in cost and complexity, this device introduced consumers, a disproportionate number of them women, to the vexing experiences we now associate with technology consumption. The struggle began when shopping for a sewing machine, trying to choose among a welter of competing versions of the technology that worked on entirely different technical principals. Once consumers decided on their purchase, they faced new hurdles, first in learning to operate the machine and how to use its tools to make adjustments and perform other operations. Sewing mechanically also meant coping with mechanical problems: diagnosing the source of trouble and then tinkering to make things right again. Finally, because this complicated new technology came with multipage owner's manuals, users became readers out of necessity, pausing in their sewing occasionally to consult the words and diagrams on the printed page for help threading their machines, winding bobbins, or installing attachments. Compared to clocks, which required relatively little new learning of their owner-users, sewing machines were far more demanding and user unfriendly.

Inventors had been trying to mechanize the making of stitches since the eighteenth century, but only in 1850 did Elias Howe introduce the first workable device for the task.[14] Neither he nor any of the other industry pioneers imagined the new technology would be used in the home; they expected sewing machines to be adopted by factories. Only late in the decade did sewing machine makers begin to envision a domestic market. In 1859 the firm started by Isaac Singer expressed this new goal in a brochure. "Public taste demanded a sewing machine for family purposes," it explained, as women desired something "of smaller size and of lighter and more elegant form; a machine decorated in the best style of art, so as to make a beautiful ornament in the parlor or boudoir." The cost of early domestic machines was high: in 1853 the price of a Howe was $125, the equivalent of many months' work for a well-paid mechanic and utterly unaffordable to poorer working-class families.[15]

In 1856 the Singer Company recognized that not enough consumers would ever have the cash to enable sewing machines to become widespread in homes, so it came up with the notion of selling them on credit, what it called "hire-purchase." Buyers would pay a few dollars down and then "hire" a machine for a small monthly amount until they had paid off the total cost plus interest. Although hire-purchase removed a major obstacle to sewing machine sales, the worsening sectional crisis and the outbreak of Civil War in 1861 retarded the market. The Union's use of machine sewing to mass-produce uniforms during the war underscored the technology's value, however, and by the end of the 1860s sewing machines were selling briskly, both to industry and to domestic buyers. The accelerating adoption of machines can be read in the increased number of patent licenses sold by the patent pool, or cartel, set up by the major sewing machine manufacturers for the purpose of sharing key patents, especially crucial early ones held by Howe. Manufacturers were required to purchase one license for each machine they built and sold, and in 1865 purchased 462,000 licenses, in 1870, 504,000, and in 1875–76, before Howe's patents expired, 577,000.[16] By the time of the nation's centennial, then, millions of American households possessed a machine for sewing, most bought on credit.

Women appear to have been initially skeptical about the technology, wondering how a machine could sew and whether it could produce seams with stitches as regular and durable as those they could make by hand. And they found it hard to believe a mere machine could do "fancy work"—that is, embroidery—like a skilled needleworker. Women also puzzled over the fact that machines of different manufacturers made stitches in different ways. Consumer bewilderment was fed by the extravagant claims manufacturers made regarding the new technology, behavior that would continue. Proponents spoke of the technology as if it were an extra "little worker" in the home, a "family friend" or assistant "lightening the labors" of women who otherwise would, "plodding along in the treadmill of care and labor, seemingly, find their work never ended."[17] If one believed the rhetoric in the fictional story "Edith Burnham, or, How the Homestead Was Saved," published by the Victor Sewing Machine Company, of Middletown, Connecticut, the technology could be a "savior," enabling a family to avert economic disaster.

The story was written by one Fannie Alden and published in 1876.[18] Exemplifying the utopian promise that would continue to be associated with personal technologies, it was a typical Victorian melodrama, its happy ending foreshadowed by its title. The protagonist, Edith Burnham, was one of the four children of a widower, Judge Burnham. The judge, along with his children and spinster sister, Aunt Hattie, had lived together in the family homestead, but shortly after the tale opens the judge, a pillar of community rectitude, suddenly dies. To the family's surprise and horror, he has left the homestead heavily encumbered with debt, which threatens not only to destroy the Burnham family's good name but also to deprive them of the roof over their heads. At this point, the youngest of the judge's children, 16-year-old Edith, hatches a plan. In the presence of her aunt and one of her brothers, she proclaims "with flashing eyes" that "the 'child' is a child no longer, Auntie, but a woman in determination. . . . I am resolved that the homestead *shall not be sold!*" The nature of her plan becomes clear a little later when Edith, with the family looking on, asks, "Isn't it a beauty, Aunty? I'm nearly wild with delight. And all my own, too!"[19]

Edith's "beauty" was her new Victor sewing machine. Although the price of the machine was $145, a sum that none of them had, Edith had struck a deal with the manufacturer; she agreed to use the machine to produce embroidery, or "ornamental work," for four years, in return for which the company would pay off the debt on the homestead. Receiving this news, Edith's siblings jubilantly praised her and her new Victor. The story ends on this upbeat and rhapsodic note, affirming the capacity of the new machine to improve a consumer's life and situation, a faith that still drives the consumption of expensive technologies.[20]

That many purchasers really could not afford the cost of sewing machines, even when sold on the installment plan, is clear from the records of the Singer Sewing Machine Company. In 1887 Singer's Philadelphia manager reported to the New York headquarters that many of his hire-purchase customers in Pennsylvania were missing monthly payments. The reasons varied but they boiled down to hard times in an economy with no safety net: in Allentown, Pennsylvania, the Lehigh and Reading Railroads were on strike; in Bradford "the shut down in the oil fields and the snow which drifted the roads so full that it was impossible to get around" undermined economic activity; in Easton, a coal strike had shut down the steel furnaces; and in West Philadelphia, consumers were in arrears due to an "almost total suspension of work in that district."[21] Not only did those who bought on credit in the nineteenth century have no unemployment compensation to tide them over in hard times, but they were taken advantage of by sewing machine salesmen. Singer salesmen were supposed to receive one commission when they sold a machine, usually 15 percent of the initial down payment, and then another, usually of 10 percent, on every payment purchasers made on their outstanding balances. But some salesmen disregarded the requirement that buyers make a down payment and sold new machines to people who paid nothing down or who traded in an old and inoperative machine, figuring that getting 10 percent on the small monthlies was better than getting nothing on no sale at all.[22]

The poorest customers were the most likely to be offered such "deals" and fall behind on their hire-purchase payments. They also were the most likely to suffer the ultimate injury of credit buying: repossession. A Singer agent responsible for sales in central Alabama, an area populated mostly by black sharecroppers like those Booker T. Washington had chided for their clock purchases, described the economic vise in which those consumers found themselves. "The section around Pike Road and Mt. Meiggs in which I have traveled the past week is in a fearful fix," he informed his superior in 1886. The entire area is "almost panic stricken" he added, and then concluded his letter by saying that "8 or 9 Machines" will have to be repossessed, the buyers "having made nothing over their rent and advances [on crops] and therefore cannot pay." The agent sympathized with his customers' plight yet worried about his own job, "well aware" that his position was "dependent upon the success and the results of the business done."[23] It is not clear what percentage of Singer sales resulted in repossessions, but figures suggest that in some areas close to one in ten purchasers eventually lost their machines due to their inability to keep up payments.[24]

Whether they paid cash or bought on credit, paid off the debt on their machine or lost it through repossession, these consumers were also the first to face the befuddling choices—in this instance, among different brands and types of sewing machines—that have become endemic to technology consumption. Early-model machines differed significantly in their construction and operation. Some produced chain stitches while others made lockstitches; some sewed with one thread, most used two threads, and a few used three. Most machines required users to feed their work forward, away from their bodies and under the needle, but the popular Willcox & Gibbs machines made operators move their work from left to right across the table and under the needle. Some machines permitted adjustments in length of stitch, while others did not, and changing the size of the stitch was easier with some than with others. Operation of most machines required working a foot treadle, while on others the user employed a hand wheel; increasingly machines had both mechanisms. Additionally, the way one wound bobbins, threaded machines, and adjusted thread tension differed from brand to brand. In short, sewing machines employed what today we might call various formats or platforms, fundamental design differences that inspired both advocates and critics. Shoppers for sewing machines had to sort out and evaluate these technical and operational differences for themselves to make a purchasing decision. The dilemma of having to choose among competing products without understanding the technology is common today, but it began with sewing machines.

Increasingly, women had the opportunity to see different machines in operation and try them out before purchase, so through trials, observation, and hearsay, they developed preferences. In the larger cities, women visited the lavish showrooms opened by major manufacturers and staffed by attractive female "demonstrators," a practice inaugurated by the Singer Company. A prospective buyer could observe these experts working the machines and, under the tutelage of a demonstrator-saleswoman, try one herself. Other opportunities to observe sewing machines in action occurred at various public competitions, which, following the marketing pattern established by makers of agricultural equipment, pitted machines of different makes against one another.[25]

During the Civil War, many Northern women also saw sewing machines at the sanitary fairs held to raise money for the Union cause.[26] Finally, many manufacturers offered women the chance to use a machine "on trial" for a few days at no cost. Some women managed to get two different manufacturers to bring machines to their home simultaneously so they could hold a competitive trial or "road test," probably with friends or neighbors from their quilting or sewing circles.[27]

Although women rarely wrote about their experiences using sewing machines, from the testimonials published by sewing machine manufacturers we can glean some sense of what aspects of the technology's performance users favored or disliked and how they assessed particular machines. Of course, companies only published favorable comments about their products, like the ones a Mrs. J. M. Sutherland gave for a Wilson Improved machine in 1872. "I have used the Wheeler & Wilson Machine five years," she says in her testimonial, "the Grover & Baker three years, and the Wilson Improved the past two years, and think the Wilson Improved much the best machine."[28] We cannot trust her judgment of best, but it is clear from even this adulatory testimonial that women engaged in a discourse of comparison, weighing and evaluating the features and performance of one machine relative to another. Putting aside testimonials that simply claimed this or that machine to be the best and focusing instead on those that allude, favorably or unfavorably, to performance criteria, we can get a reliable sense of women's preferences regarding the new technology and the problems they sought to avoid.

Women, the testimonials suggest, valued machines that did not require great physical effort to operate. The claim of Mrs. E. R. Boissat, of Alexandria, Louisiana, about her Grover & Baker machine, that a 10-year-old girl "can sew on it with the greatest ease," evidences women's desire for machines that were "easy running."[29] Other testimonials indicate that women also favored mechanical simplicity. Typical is the claim of Mary

McMully, of Log Harbor, Long Island, about a Wilson machine. "One hour's instruction will serve for learning its simple construction and easy working," she wrote, a point echoed by the testimonial of J. Stokes, of St. Louis, on behalf of his wife, who "had no experience [in sewing] whatever, and no particular aptitude as a Machinist" yet who nevertheless readily mastered a Wheeler & Wilson.[30] Similar was the 1869 testimonial of Mrs. E. M. Benjamin of Riverhead, Long Island: "When I was first married," she wrote, "my brother sent me a Grover & Baker, and afterwards a Wheeler & Wilson; but I did not like the stitch of the first, the other was too complex, and both were too noisy to suit me." Eventually she bought a Willcox & Gibbs, which "has worked to a charm."

Finally, sewing machine testimonials attest to the desires of users, like technology consumers ever since, for machines that do not give trouble and break down. Women wanted technology that stayed healthy, as is evident in the testimonial of Mrs. N. Mills, of Lakeville, California. She had used her Willcox & Gibbs "nearly nine years" and claimed "it has no 'spells'—is never out of order."[31] Another Willcox & Gibbs user said she was "often surprised at the complaint ladies make—of their machines (other kinds) having 'fits.' Mine is never out of order."[32] Anthropomorphizing a sewing machine as if it were alive and capable of fits, spells, or willful behavior, women revealed what would become a familiar consumer response to complicated personal technologies. Their comments expressed both the unavoidable intimacy users developed with the workings of their machines as well as the value they placed in a machine's good "behavior"—and, by implication, their desire to avoid the exasperation they might feel when it threw a fit or otherwise misbehaved.

Along with the technical details involved in choosing what sewing machine to buy, purchasers also confronted a problem that had little preindustrial precedent. This was whether the machine offered for sale was really "new" and whether it was made by the firm whose brand name it bore. The Singer Sewing Machine Company, the nation's largest producer, claimed that fakes carrying the Singer name were "doing much injury" to its business.[33] Trademark laws were new and weak, and they were barely enforced, leaving consumers to cope with the presence of bogus machines on the market. What exactly bogus meant was hard to determine, as some manufacturers and retailers, especially after the major sewing machine patents expired in 1877, believed that "any person has as much right to make the genuine Singer sewing machine as the Singer Company itself."[34] Also confusing was the fact most companies reconditioned old machines and then sold them as new. Even Singer's own salesmen sold old demonstrators as if they were new machines. Although the Singer *Canvassers' Manual* warned that "nothing has so bad an effect on the chance of a sale, with the generality of people, as the suspicion, even, that the machine is not entirely new," no clear standards had been yet established as to what exactly "entirely new" meant.[35] Did it mean a machine that literally had never been run? What if the machine had been used just a few times as a demonstrator? And what about one that had been used for nearly a year, sent out on trial to numerous prospects, and never sold—was it also "entirely new"? That the industry had not yet resolved the ambiguity surrounding the meaning of "new" or addressed the dubious ethics of selling old, reconditioned machines as new, is suggested by another piece of advice in the manual. Salesmen who were reclaiming a machine that had been out on trial were implored to "see that the machine is thoroughly cleaned up, and every trace of its having been used is removed."[36] The cleaned-up machine would continue to go out on trial to home after home until, eventually, it was purchased—as a "wholly new" Singer.

Consumers already had a low opinion of door-to-door salesmen when sewing machines began to be peddled in the 1860s and 1870s. More than two generations of peddlers, including clock sellers, had left negative stereotypes. Peddlers evoked either the image of a shrewd, calculating Yankee, pushing fake nutmegs carved out of ordinary wood and other shoddy wares, or a tightfisted Jew, haggling over the last penny to make a deal. As the leading retailer of sewing machines and the company most reliant on door-to-door sales, Singer tried to avoid the opprobrium associated with peddlers by referring to its agents as "canvassers," but Singer's competitors called them on this ruse. L. L. Richmond, a St. Louis retailer who sold reconditioned Singers and other machines from his downtown store, attacked Singer by invoking the image of the stigmatized peddler. In a broadside with the single word *STOP* set in gigantic type at the top, Richmond recounted a story about how the Singer Company had sued a customer who bought one of its machines. The buyer had purchased from "peddlers," Richmond asserted, who then absconded with the money without remanding Singer's share, at which point the manufacturer went after the customer in court. In concluding its anti-peddler pitch, Richmond posed a question intended to prove worrisome to any potential customer: "Query—Can't you save both [money and a lawsuit] by dealing with a *Responsible* merchant instead of an irresponsible Peddler. You pay your money, take your choice." Richmond's broadside concluded by noting he was the "sole agent for the CHICAGO, SINGER, VICTOR, DAUNTLESS, REMINGTON & NEW WEED SEWING MACHINES," along with being a special agent for "Singer, New American, Automatic, St. John, White, Wilson, Davis, Whitney, New Home, Domestic, Grover & Baker, Florence, Wheeler & Wilson, Howe, Crown and Wardwell."[37] His handbill underlined the perils of shopping for a technology that came in a dizzying variety of brands and types of machines and which required consumers to choose between "responsible merchants," selling out of established, brick-and-mortar premises, and "irresponsible peddlers," here today but gone tomorrow.

Even in palatial urban sewing machine showrooms, presumably protected from bogus machines and peddlers, shoppers needed to cultivate the attitude of caveat emptor, or "buyer beware." The ornate furnishings and well-dressed saleswomen of such stores were intended to make middle-class women feel at home and to mask the calculated pressure placed on prospects to buy. One of Singer's showroom sales managers in an eastern city, a "Mrs. G.," explained some of the psychology used to break down consumer resistance.[38] "I study human nature," Mrs. G. had told her boss, who had asked how she sold so many machines. "I am very careful to see that the office is as 'clean as a pin,' that the show windows are cleaned often," she added. Another detail she insisted upon was that her Singer showroom window always contain "a machine and samples of work," so that even "disinterested passers by will be attracted." When customers entered her showroom, she did not "run down" the machines of competitors but instead talked about the ways Singers were "superior." Indeed, she helped make the machines superior through special tuning, somewhat the way "factory" stock cars are tweaked to outperform models right off the showroom floor. "I am very particular if I send a machine out on trial against a competitor," Mrs. G. explained, "to go out in the shop and see it myself. See that the tensions are perfect and try each attachment." When demonstrating machines to prospects in the store, she also followed a script. "If I can make a sale without showing all the attachments I do so." But if a customer was "interested in fancy work" or embroidery, Mrs. G. went through a routine. Preferring to "leave the ruffler until the last (except the embroiderer) and as the customer becomes enthusiastic over this attachment I reach the climax by showing off the embroiderer." As an experienced operator, and using carefully set-up and tuned machines, Mrs. G. achieved the desired climax with regularity, selling buyers on the belief that they, too, could achieve the same results when they got their Singers home.

A nineteenth-century trade card depicts the exciting moment when a purchased sewing machine is delivered to the home—delivery usually a necessity because the machine and its stand or cabinet were not easily carried. A Wheeler & Wilson wagon has pulled up in front of a house and the whole family has gathered on the front steps to watch as the machine is unloaded and carried toward the door.[39] Once inside, the new technology probably evoked the same admiration—it's "a beauty"—as had Edith Burnham's new Victor, for early sewing machines were visually stunning. They were usually painted with glossy black enamel and embellished with gold striping and the maker's name, spelled out in ornately decorated letters also in gold. Other parts were nickel plated and shone brightly. Altogether, the sewing machine's polished, all-metal construction anticipated the design of many subsequent manufactured devices like bicycles, automobiles, and eventually consumer electronics. Indeed, part of the seduction of a sewing machine was its new and shiny industrial aesthetic.

To make sure sewing machines appealed to feminine sensibilities, most manufacturers added delicate painted flowers to the ornate striping and lettering, giving the machines the look and allure of lacquered Japanese boxes, which were the rage at the time. Domestic sewing machines were also physically dainty. They adhered to Singer's 1859 dictate that home machines be "smaller" than industrial ones and appropriately scaled to women's more diminutive hands, although as the century wore on and women fully embraced the idea of sewing mechanically, the size of home and commercial machines tended to converge toward similar dimensions. Even if early domestic machines were smaller, they were invariably attached to a stand or cabinet incorporating the foot treadle, making the overall unit quite sizeable and heavy. Only when electric motors began to replace treadle power in the early twentieth century did sewing machines finally became "portable," though they remained heavy.

Having chosen a machine and had it delivered and set up in her parlor or dining room, a woman confronted the challenges of learning to operate the device. Victorian women grew up handling needle and thread and would have been intimately familiar with sewing, including darning, mending, and fancy work such as embroidery and needlepoint. They quickly discovered, however, that even though a sewing machine employed needle and thread it was an unfamiliar, even alien entity. To sew mechanically required women to manipulate a complex interface involving the levers, knobs, screws, belts, wheels, and treadles of their machines. Some women just threw up their hands at the prospect of mastering the complicated new technology. As the 1876 "Edith Burnham" story suggests, youth tended to adapt more rapidly to the sewing machine than their elders, just as would be the case with later technologies. Shortly after Edith brought home her new Victor and saved the homestead, her older sister, Grace, perhaps in her late teens, also demanded to "learn" the machine. Edith agreed and instructed Grace on the rudiments and allowed her to practice with the Victor. As the judge's sister, Aunt Hattie, watched Grace and Edith easily handle the sewing machine, she exclaimed: "Plate-Gauge, Ruffler, Gatherer, Shuttles, Bobbins, Needles, Wrench, Screw Drivers, Oil Can, Throat-Plate, Bottle of Oil, Instruction Book, it almost took my breath away to hear Grace rattle off their names as glibly as if she had been familiar with them all her life."[40] Her wonder and exasperation typifying the manner in which older generations seem always to have responded to recently introduced technology, Aunt Hattie could not imagine herself ever learning the new machine.

Beyond becoming familiar with the numerous components described by the technical jargon that so intimidated Aunt Hattie, what exactly did women have to learn to sew mechanically? "The operator should first become familiar with the treadle motion," advised a typical instruction booklet, referring to the foot treadle used to power the mechanism producing the sewing machine's stitches.[41] Although for centuries treadles had driven saws, lathes, grinders, and other machinery in factories and workshops, few women would have seen, let alone used, one before the device was adapted to sewing machines in the 1850s.[42] Now women had no choice but to master treadle operation. To do so often meant abandoning any favorite armchair women might have used while toiling away the hours doing needlework; now they had to sit in a straight-backed, armless chair pulled up close in front of the machine so that their knees went under the table with their feet positioned squarely atop the treadle. Once suitably positioned, an operator would push down on the treadle, first with her toes and then with her heels, thereby rocking it back and forth. This oscillating motion would be translated into rotary motion by a pitman, or connecting rod, and crank so as to turn a wheel that, via belt and pulley, drove the mechanism. The up-and-down action of the needle, the movement of the feed that pulled cloth under and through the needle, the coordinated movement of the shuttle holding the second thread, and other elements of the machine all depended on proper "treadling."

Using the treadle was tricky: "practice is necessary in order to give a steady and uniform revolution to the wheel," one manual advised.[43] Novices invariably struggled to get the wheel to uniformly turn in the same direction, that is forward, for the treadle crank could rotate either forward or backward.[44] Many early sewing machines sewed only when going forward, that is, with the feed mechanism pulling the cloth away from the operator and under the needle. If the treadle operator caused the machine to go backward, so to speak, the feed would back up the cloth but would not produce stitches. A second problem was controlling speed, as it took a practiced touch to sew slowly with a treadle machine. Paradoxically, while speed was a major selling point for sewing machines, operators often wanted to slow down—while turning a corner to make a hem, say, or when sewing repeatedly over a single location to secure the thread prior to cutting it. If a machine had a hand wheel along with a treadle, as eventually all of them did, seamstresses would turn the wheel manually to advance their work slowly or back it off; with a treadle alone, as was the case on some early sewing machines, controlling direction while running slowly was difficult. All of these operations required time-consuming practice and considerable patience.

In learning their complicated machines, women had the assistance of printed instructions. Sewing machine makers commonly included booklets of twenty to fifty pages with their products, texts that a student of technical literature, John Brockmann, claims were the first owner's manuals.[45] They included much more information than the brief instructional labels, containing only a couple of paragraphs, that came with mass-produced clocks in the early decades of the century. The authors of this new genre of consumer literature were anonymous, as have been the writers of virtually all such literature, from the days of early clock instruction labels to today, but the owner's manuals they wrote represented the industry's major response to the complexity of personal technologies; manuals were essential to mass consumer adoption of such machines. Titles like *The Howe Sewing Machine Instructor* suggest how the industry considered the manual a surrogate for a human tutor.[46] Even when a purchaser received initial instruction from a demonstrator or salesperson, her printed "instructor" became the only source of information and help once she was home alone with her machine. "This book should be kept for reference," the Wheeler and Wilson manual reminded buyers.[47] The importance of these instructional texts is underlined by the 1888 plea of a Milwaukee Singer agent to his boss in New York. "It is quite necessary," the agent wrote, "that we have Norwegian Instruction Books for the V.S. No 2 [machine], as a large number of our customers in both Minnesota and Dakota speak nothing but that language."[48] Technology consumers obviously needed help, and they needed it in a language they could understand.

Sewing machine manuals not only described but, unlike clock directions, also explained and, even more importantly, visualized things for readers, employing a variety of graphics in support of the text. The "profuseness" of illustration in sewing machine manuals, as Brockmann has shown, exceeded that in publications for earlier technical products and sometimes even those of later consumer technologies like the automobile.[49] Most sewing machine instructors included a general illustration of the overall machine, a "View of the Machine Ready for Work" or a "Transparent View in Perspective."[50] These functioned as maps to orient women to their new machine and its many parts—*cross-head, Pitman, brass ferrule, eccentric, connecting rod, gauge, flywheel, set screw*—the terms were printed on the illustration or listed in a numbered key that referenced numbers on the illustration.[51] The "Transparent View" is unusual in that the artist has rendered the machine as if he had X-ray vision, enabling the reader-viewer to see interior elements that in reality would be hidden by other parts. These manuals also commonly included another kind of graphic, hands-on illustrations.[52] These are drawings or engravings depicting a pair of hands working with the machine. The hands may be setting or installing a needle, using an attachment like the hemmer, or performing some other task. Usually the hands were drawn from an over-the-shoulder view as if they were those of the operator, a form of representation that gave the illustration greater instructional and psychological power. Women who in effect saw their "own" hands working the machine in the manual would have found it easier to learn to use the technology and feel comfortable sewing mechanically.

An illustration usually addressed the threading of the machine, as did the one depicting it ready for work. It showed women the complicated path, from spool to needle, they had to take with the upper thread of the machine—through eyelets, around hooks, and eventually through the needle's eye itself. Remembering this path was hard, and so a woman who had not used her machine in a while might need to consult the manual for assistance. The needles for machine sewing were different from those employed in handwork, being thicker and having their eye near the point rather than at the other, blunt end. The nonpointed ends of sewing machine needles were shaped so as to fit tightly into the oscillating heads of machines and usually had to be oriented precisely ("It will be observed that the double grooved needles have one groove which runs below the eye of needle. When using these needles be particular to set the needle with the longest groove to the left.")[53] Additionally, many needles only fit a particular make, and sometimes even model, machine. When a needle bent or broke, a not unusual occurrence due to the force with which the machine drove it through the fabric, women might find it hard to find the correct replacement. The resident of a small Delaware town in 1864 was exasperated by this problem and wrote Singer requesting a dollar's "worth in needles, assorted sizes, for one of your letter A machines. Send by return of mail," she implored, "as I have not a single needle to use." Perhaps because of previous experience, she added this warning: "in packing the needles in the envelope see that they are not placed in the part that the P.M. will likely stamp the letter on."[54] As a consumer she had discovered that technologies could have prodigious appetites for provisions and replacement parts and be very fussy as to what kinds they would consume.

Having the proper sized needle correctly "set" and threaded was not the end of a woman's preparatory work for sewing mechanically, as machines usually employed two threads, a lower as well as an upper one. Handling the second or lower thread was more unusual, in that operators first had to wind it onto a bobbin similar to those used on looms, although much smaller. This bobbin then was inserted in a shuttle beneath the machine's worktable, and the threading procedure was completed by drawing the loose end of thread from the bobbin up and through the loop made in the upper thread by the action of the needle. After these extensive preparations, the operator was finally ready, as a Howe manual put it, "to commence sewing."[55]

Unlike sewing by hand, however, sewing mechanically required that women frequently stop to manipulate the machine's controls or make adjustments. They learned to lift and lower the presser foot, usually by moving a lever, that held cloth firmly down against the worktable, fed it forward under the needle, and made the stitches. Only if they raised the foot could operators maneuver their work under the needle, turning it or lifting it. They raised the presser foot before starting or ending a seam and when turning a tight corner. Like the clutch pedal of a car with manual transmission, the presser foot was always in use, an essential operation one performed in running the machine. Operators also had to occasionally stop and adjust the tension of one or another of the two threads used on their machines. Both had to be equally tensioned, otherwise the stitches either on the top or bottom of the fabric would be tighter, causing it to pucker, a problem that never occurred when hand sewing with a single thread. The means for adjusting the tension varied machine to machine, but usually operators tightened or loosened a nut or turned a thumbscrew, both of which essentially squeezed the thread on its way from spool or bobbin to the needle. Operators learned that numerous factors might affect thread tension: the type of thread itself, that is its weight and thickness or whether it was coated so as to make it more slippery; the nature of the fabric being sewn and the number of layers; the length of stitches being made; and even the humidity, for drier thread would slip more easily through the machine. Manufacturers sought to improve and simplify the way their machines controlled thread tension, and some, like Willcox & Gibbs, claimed their machines had "automatic tension" requiring no adjustments.[56] The reality, however, was that sewing mechanically meant periodic tinkering with thread tension, one of the several additional tasks and interruptions women endured when sewing with machines.

These and other operations often required women to use the tools that sewing machine manufacturers provided them, usually a screwdriver, wrench, and oil can. Such tools would hardly have been familiar in most mid-nineteenth-century households, especially among city dwellers, and not only women but also many men had probably never used them.[57] Wrenches and screwdrivers were uncommon because screws, nuts, and bolts were not yet widely used in the manufacture of consumer goods. Mass-produced clocks did not use them, and furniture was held together by joinery and glue, while the parts of another common consumer device after the Civil War, the cast-iron kitchen stove, were interlocked together like a large jigsaw puzzle, without any fasteners.[58] The basic fixed-end, open wrench provided with sewing machines was so alien in 1875 that Willcox & Gibbs felt it necessary to include an engraving of the tool, captioned "The Wrench," in its owner's manual and to explain that to use it to set the needle one must "place the wrench on the 'nut'"—also labeled— and "turn it to the right until the needle is firmly held in its place."[59] The prospect of having to use strange new tools may explain some women's reluctance to adopt sewing machines, an anxiety obliquely hinted at in a testimonial praising a Grover & Baker machine, whose owner claimed she used it constantly for "eleven years and never loosened a screw."[60] That must have been an exaggeration, for screwdrivers or wrenches were needed to set or change a needle, install or remove attachments, and on some machines even to adjust the length of stitches.

A remark by the fictional Aunt Hattie in the "Edith Burnham" story captured another side of the tool-using reality of sewing by machine. "There wasn't a screw in that sewing machine of which Edith was not mistress," Aunt Hattie proudly says of her niece.[61] Understanding and working with the screws and bolts of their sewing machines, women were acquiring a new kind of technical competence and learning to be consumers in an industrial society. Some women so enjoyed this new role of machine "mistress" that they provoked the ire of manufacturers. "Purchasers sometimes say," a Howe instruction book reported, " 'Why, I took my machine *all to pieces*, and put it together again, soon as I got it.' No worse thing could be done." It then sternly warned readers: "Don't attempt to tinker the machine."[62] Clearly, the men in the industry did not consider the aggressive disassembly of their products to be suitable behavior for women, but theirs was not simply a stereotypical or misogynist response.[63] Sewing machine makers, like manufacturers of automobiles and other personal technologies later, legitimately worried that tinkering consumers might upset the delicate working of products. They therefore sought through warnings and warranty provisions to prevent such behavior, hoping thereby to increase customer satisfaction.

Even if consumers did not take their machines "all to pieces," they on occasion needed to remove and replace a worn or damaged part, even if only a broken needle. The sewing machine was one of the first consumer products for which manufacturers kept an inventory of spare or replacement parts. Many companies in fact provided buyers with a complete parts list in the instruction manual, often illustrated with the graphics linked to part numbers, so consumers could identify what part they needed and order it even if they did not know what it was called.[64] In ordering parts consumers usually had to provide the model of their sewing machine as well as other information such as its size, serial number, and year of manufacture, all of which might or might not be stamped visibly on the machine itself. A replacement part, being interchangeable, would fit and work only if it was the correct one for the consumer's particular machine. The need for precision in ordering was a consequence of precision manufacture, but consumers found it a difficult lesson to learn.[65] One who presumably was not satisfied was the Attica, Ohio, woman who in 1865 wrote the Singer Sewing Machine Company asking it to send her "1 small bar or hook like drawing enclosed."[66] Such lack of specificity was fatal and would not produce satisfaction; technology consumers would continue to find the problem of parts replacement challenging until they learned to be precise about what they needed.

As for repairs and maintenance, sewing machines seem to have been quite sturdy and reliable performers, and consumers generally needed only to replace needles, the rubber belts that transmitted power from treadle to mechanism, and other minor components. Indeed, many sewing machines provided decades of service without any repairs or service, a fact attested to by the large number of still workable treadle units surviving from a century or more ago. The technology did, however, require occasional cleaning, as lint and dust easily penetrated its open mechanisms, and the presence of oil on many parts facilitated the buildup of grime. Lubrication, however, was probably the most frequent maintenance task owners performed. "Oil is needed wherever a moving surface comes in contact with another part," explained one manual.[67] It was hard to know just how much oil to apply, however. The *Directions for Using the "Domes-*

tic" Sewing Machine instructed women that "when in use, the machine should be oiled from once to twice a-day," while the *Directions for Setting Up and Using the Union Button-Hole Sewing-Machine* urged users to oil "*All* the parts where there is any friction, except around the clamp and plate," once per day, while "the cam which works against the top end of nipper lever" should "be oiled three or four times per day."[68] "Only a few drops" at a time, suggested another manual, lest "the machine become gummy and run heavy," in which case owners should clean its mechanism with turpentine or kerosene.[69] Further complicating things, some machines had components that users were warned not to lubricate. "CAUTION," the Wheeler & Wilson manual boldly stated: "NEVER OIL THE TENSION-PULLEY."[70] In short users had to discover for themselves what constituted "too much oil" versus "dryness," and how dry their particular machine had to be before it ran "heavy."[71] Manufacturers, wanting their products to run well, and in many cases selling oil on the side, pushed oil consumption. The Wilson Shuttle Machine Company provided buyers of its machines with a little vial of the firm's "No. 1 quality of oil, manufactured especially for our trade, and put up by ourselves, with our name blown in the bottle," and encouraged owners when they ran out to buy more of the same.[72] Wilson didn't produce oil itself but bought it in bulk and then repackaged it for sale, no doubt at a nice profit. Consumers were rightly confused as to whether all oils were in fact the same and had no way of judging quality.

Whatever confusion and challenge women experienced with regard to sewing machines, they continued to adopt them in increasing numbers. After the original Howe patents expired in 1877, machines became more affordable than ever as manufacturers no longer had to pay a license fee to the patent pool for every unit sold. As the years passed and engineers and manufacturers gained experience designing and making them, machines also were better made, more reliable, and easier to use. Changes in marketing also spurred adoption, for starting in the 1880s mail-order firms like Montgomery Ward and Sears and Roebuck began selling sewing machines through catalogs, offering low prices with money-back guarantees. The technology's appeal also benefited from the widespread sale of inexpensive paper patterns for dresses, skirts, shirts, and other garments, all offered in standardized sizes. It became easier for women to imagine themselves as do-it-yourself seamstresses, outfitting themselves and their families stylishly and inexpensively.[73] Reflecting on women's embrace of sewing machines, in 1885 the editor of the trade paper *Sewing Machine News* found it remarkable that, in barely a generation, a radically new invention had become a fixture in the lives of ordinary people everywhere. "The ignorance, widespread among all classes, of sewing machinery and its management" knew no bounds a generation ago, he observed, yet now "so general is knowledge of sewing machines that very many women understand how to operate them though they have never owned them." Just thirty-five years after its first introduction, he concluded, "the sewing machine is no longer a mechanical mystery."[74]

SCHOLARS CALL THE PROCESS WHEREBY CONSUMERS ACQUIRE, become familiar with a new technology, and no longer view it as mysterious "diffusion." The term analogizes the spread of artifacts in a society to the behavior of a gas, liquid, or soluble compound in another medium—for example, to the way ammonia fumes rapidly disperse throughout the air in a room. But while statistics as to the number of parts of ammonia to oxygen in a room may tell the whole story of physical diffusion, statistics on the number of sewing machines, radios, automobiles, cell phones, or any other device sold per capita tell us little about the attitudes or experiences of those individuals who adopted and used such technologies. The language of diffusion when applied to technology obscures the subjective stories of struggle by consumers or even resistance to adoption. Some students of technology-society relationships therefore prefer to describe the process of adopting new technologies as one of "domestication" rather than diffusion, the folksier term better accounting for the variations in experience people have in bringing new machines into their homes and lives. Thinking of this process as domestication leaves room for the common tendency of older individuals to be more resistant to or even intimidated by new technologies, while youths learn them without struggle. We glimpsed this already in the nineteenth-century story about Elizabeth Burnham and her younger sister, both of whom learned the sewing machine quickly while their older aunt Hattie simply could not imagine herself doing so. Over time, however, such generational divides usually fade, and many who initially hold back or resist come around to adopt the technology. It is the young who ensure the domestication of a new technology, then, for those who grow up with it seldom experience its mystery and newness as their elders do; they take the technology for granted as a natural part of their world.

By the time sewing machines became commonplace, Americans were being enticed by a growing torrent of mass-produced machines. Some of these new devices aimed at saving labor, time, or effort: carpet sweepers, washing machines, lawn mowers, and meat grinders all fell into that category. Others, such as bicycles, hand-cranked music players, telephones, electric heaters, and snapshot cameras, gave buyers new and marvelous capabilities and powers.[75] At the time, consumers did not lump these devices together under the rubric of "technology," of course, and in fact early adopters of each of these machines faced unique and different challenges. Some technologies were more difficult to learn to operate and use than others or required unusual knowledge or practice in order for one to become proficient operating them. In some instances consumers had been prepared by experience with earlier devices for the new technology, but when a technology was wholly novel and buyers brought little relevant past experience to their encounter with the new device they were likely to consider it more challenging. And while no machine is foolproof, the kinds and frequency of troubles people had with different machines varied considerably. As we've seen, some, such as clocks, were rarely problematic, while others, such as sewing machines, demanded occasional tinkering or adjustment. Still others, as we shall see when we take up the early automobile, had complicated innards that required continued hands-on intervention.

In concluding this examination of nineteenth-century technologies, I'll briefly compare three devices that illustrate the different levels of challenge faced by new adopters: the bicycle, telephone, and snapshot camera.

The bicycle as we know it, the so-called "safety," appeared in the late 1880s and early 1890s and replaced the earlier Victorian version of the bicycle, the highwheeler, or "ordinary." It was also a two-wheeler but had a gigantic front wheel that measured three feet or more in diameter. As should be obvious, mounting a highwheeler demanded a gymnast's agility along with considerable courage, for the rider's seat was atop the big front wheel. These cycles seldom had brakes and no freewheeling, that is, the pedals were fixed to the front wheel and spun around with it. "Headers"—when riders were thrown over the handlebars—were common, so only young men and boys tended to ride these machines. The safety bicycle, however, attracted young and old riders of both sexes. Users sat between the two wheels atop the frame, able to touch the ground with their feet. Pedaling they drove the rear wheel via a chain that, even without gearing, provided a mechanical advantage which took much of the arduousness out of cycling. Safety bikes with freewheeling enabled riders to stop pedaling and coast, the toothed sprocket on the rear wheel able to remain stationery while the wheel itself turned; pushing backward on the pedals, however, engaged the coaster brake, permitting riders to easily stop. During the mid-1890s, millions of Americans were seduced by this new form of mobility, a technology that allowed them to pedal away from city crowds into the countryside or escape tight parental supervision or other chaperoning. They giddily relished the freedom and exhilaration of what historians have called the decade's "bicycle boom."[76]

Except for what they knew of local geography, first-time cyclists brought little useful information or prior experience to their encounter with this new form of personal mobility. Although the controls, or interface, of a "wheel," as a bicycle was often called, were simple enough, consisting of the handlebars for steering and the pedals for propulsion and braking, first-time users were challenged by having to learn to balance on its two wheels. For adults in the 1890s, this could be a painful, protracted, and embarrassing matter. Consider the experience of Francis Willard, the 53–year-old suffragist. It took "about three months," the self-described "sedentary" recalled, "with an average of fifteen minutes practice daily, to learn, first, to pedal; second, to turn; third, to dismount; and fourth, to mount independently this most mysterious animal."[77] Like others, she compared the bicycle to the familiar horse, but relative to a horse a wheel was not only mysterious but also stupid. Willard believed it took more knowledge and skill to ride one than to ride a horse. The author of the 1896 handbook *Bicycling for Ladies*, Maria E. Ward, agreed. "If you are an equestrian," she wrote, acknowledging the intelligence of horses, "you will meet with many unexpected problems" that with the animal's help plus your own experience can be solved. But "the bicycle will do nothing for you," and its "lack of horse-sense must be supplied by your own intelligence." A wheel was "unable even to stand by itself," she added; "you cannot teach it anything and there is really much for you to learn."[78]

The learning Ward had in mind extended beyond just mounting and dismounting, pedaling and braking, and steering and balancing. It also included how to adjust and repair a bicycle. Wheels in the 1890s were less robust than modern bikes, and cyclists had to be prepared when on the road to replace broken wheel spokes and chains and to patch leaky or punctured inner tubes. A "tool bag," explained cycling handbook author H. C. Cushing, "is always furnished with your wheel free of charge." It usually included a tire repair kit, pocket air pump, monkey wrench, and what one manufacturer, Telegram Bicycles, called a "Telegram tangent spoke"—a device to remove or tighten wheel spokes, which often got loose or broke.[79] "Never leave your tool bag at home," warned Cushing, "as your wheel seldom needs any attention when you have it with you."[80]

The bicycle components requiring the most attention in the 1890s were pneumatic tires. Unlike the machine's gears and cogged wheels, its nuts and bolts, and the screwdrivers and wrenches used to deal with those parts of the machine, the bicycle's rubber, inflatable tires constituted a new and unfamiliar technology. The tools and techniques needed to repair pneumatic tires were novel, and so the tire repair kits included with new bicycles came with printed instructions, explaining to users how to remove the wheel and inner tube, prepare a rubber patch, affix it to the inner tube with a special kind of adhesive, and then replace the tube in the tire and put it back on the wheel before inflating. While wheelwomen and often wheelmen's previous experience with sewing machines was of no help in coping with tire trouble, what the bicycle taught them about flats, blowouts, and their repair would serve them in good stead a few years later when many became early adopters of horseless carriages.

Like the bicycle, the telephone also was a radically new late-nineteenth-century technology, one that miraculously enabled users to converse with people miles away and out of sight. Although the first phones were installed in heavily trafficked places like railroad stations, drugstores, or hardware stores, they rapidly entered the home.[81] Early adopters struggled with the new technology, not because it was difficult to operate but because of how it upset established Victorian behavioral conventions. As historian Carolyn Marvin has shown, the genteel classes were shocked that utter strangers could electrically intrude on the sanctity of the family and that men could by phone virtually call on women to whom they had never been introduced. Both were affronts to social propriety. Victorians also struggled to devise acceptable telephone etiquette, what to say when announcing themselves over the telephone on placing a call and, even trickier as they might be speaking to total strangers, what to say on first answering the phone.[82]

Operating the new communications technology, however, proved easy, literally child's play in that kids learned to do it. Users did need to maintain the wet cell batteries that powered early telephones, a messy, somewhat exacting, and potentially hazardous undertaking, but placing a call simply meant turning a crank on the instrument to ring "Central," as the switchboard operator was called, and then asking to be connected to the desired party; taking a call was even less demanding, in that one only had to lift the earpiece and talk. To be sure, users seldom had the slightest clue how telephones worked; the underlying electromagnetic principles were abstruse and hard to understand without formal schooling in physics and higher mathematics. As a result, users were generally incapable of diagnosing or fixing technical troubles with their phone or with the larger system of which it was a part. Because telephone companies invariably owned the phones and assumed the responsibility of coping with technical problems, users were not put off by the technology's complexity.

The camera offers an even more dramatic example of how a technology based on hard to understand science, in this case chemistry, could be transformed into a user-friendly consumer product. Before George Eastman introduced the first snapshot camera in 1888, making photographs was an arduous and complicated process. In a darkroom utilizing chemicals they prepared, photographers first had to prepare the light-sensitive copper or glass plates on which an image would be recorded. After exposing these plates in a camera, that is, taking pictures, they returned to the darkroom and developed or "fixed" the images to preserve them. Because taking pictures was intimately wedded to the use of chemistry and darkrooms, most photographers were professionals, although some men and a surprisingly large number of women became accomplished amateurs.

Eastman's innovation in 1888 was to separate the act of taking pictures from the time-consuming and intellectually demanding labor of preparing chemicals and using darkrooms. To do this he devised a new kind of camera that used rolls of flexible celluloid film instead of the fragile glass plates that had to be processed immediately after exposure. Buyers of a Kodak, the name he deliberately chose for his cameras because it meant nothing good or bad in any language, got a camera already loaded with enough film for one hundred pictures. After shooting their pictures, users sent the entire camera to Rochester, New York, where the final step in Eastman's innovation involved legions of female darkroom workers who extracted the roll of film, developed and printed it, and then reloaded a new roll before shipping the camera back to the customer along with the printed photographs. Users were totally freed from any involvement with the inner workings of their cameras and even from handling film. Kodak customers benefited from the recently introduced fourth-class U.S. postal rate that enabled them to inexpensively send their cameras through the mails. Because this revolutionary new means of photography allowed a user to take pictures at any moment without any preparation, they came to be known as snapshots.

The slogan Eastman devised to market the Kodak, "You Press the Button, We Do the Rest," brilliantly underscored the revolution his system had brought to photography, but it also captured the essence of a user-friendly technology. It's hard to imagine a simpler interface than its single push-button control.

Snapshot takers would soon learn that even with this most user-friendly technology there was more to successful photography than pressing a button. To get satisfactory pictures, they should not point their cameras directly into the sun or shoot in dim light or total darkness without a flash. Users also learned that holding the camera steady was essential if they wished to avoid blurring their photos. Equipment trouble was seldom an issue, however, as the only moving parts on an early Kodak were its shutter and a simple mechanism to advance and rewind the film (later models would add simple aperture controls), and these seldom needed maintenance or repair. The snapshooter's freedom from having to deal with film, however, quickly ended. Soon after launching his first Kodak, Eastman's Rochester film-handling facilities were overwhelmed with the flood of cameras coming in from all over the country. Although his inventive simplified system freed users from the need to handle delicate film, it was inefficient to send cameras through the mails to his factory essentially as containers for film. Eastman came up with the idea of prepackaging film in small, light-protective canisters so owners could load and unload their own film. This innovation permitted the decentralizing of processing, as owners took their exposed rolls to local laboratories for development and printing, essentially the system still in existence today.

Over the years, camera makers introduced different sizes and formats of cameras, which led to the proliferation of film types. The work of owning even a snapshot camera—in particular, keeping it supplied with the proper-sized film—became more complex, as few retailers stocked all sizes of film and users, especially when traveling, had to plan ahead in order to have enough of the right film. As with other personal technologies that came in a variety of makes, models, sizes, and formats, the parts or supplies that worked for some cameras were not suitable for others. Having the wrong-sized roll of film, like having the wrong needle for a sewing machine or incorrect-sized tube for a bicycle tire, was no better than having none at all.

By the time the twentieth century arrived, Americans generally understood this. After a century of technology consumption, they had owned and used a range of manufactured devices, from mechanical clocks to sewing machines, snapshot cameras, bicycles, and phonographs. From their experience with machines, they had acquired a rudimentary mechanical education and level of technical literacy superior to citizens of most other countries. To be sure, not every individual possessed the same degree of technical knowledge; it varied from person to person and group to group. Yet people were not excluded from technology consumption by race, class, or gender, and even black sharecroppers owned clocks and sewing machines. Furthermore, while people who were wealthy, white, and male enjoyed disproportionate benefits in society, those characteristics by no means insulated them from the confusions, frustrations, and struggles accompanying the adoption of new technologies. Machines tended to treat all consumers equally. And though men invented, designed, and built most of those machines, and as a group monopolized much knowledge about technology, many women, as we have already seen, became knowledgeable about sewing machines, bicycles, and other devices in their roles as consumers and users.

Almost everybody would at times feel stupid and ignorant in the presence of a new personal technology emerging at the turn of the century, the horseless carriage. By far the most complex machine ever produced for the consumer market, the automobile would have many times more parts than any clock, bicycle, or sewing machine. In learning to run an automobile, operators had to coordinate hands and feet as they did when using a sewing machine or riding a bicycle—but do so while moving at speeds of twenty to thirty miles an hour. Additionally, unlike earlier consumer technologies that were wholly mechanical, the automobile had components and systems that were chemical or electromagnetic in their operation, requiring motorists to pick up previously alien scientific knowledge. Automobile owners acquired a familiarity with the chemical fuel used by a vehicle, whether gasoline or kerosene, as well as the variety of lubricants required, including greases as well as oil; they learned that hydrochloric acid made their batteries work and calcium carbide generated the acetylene gas used in headlights. From their ignition systems, motorists learned something about short circuits, sparks, and voltage. And when their car gave them trouble, diagnosing the cause and tinkering the machine into running again taught owners that the early automobile was the antithesis of a user-friendly, you-press-the-button-and-we-do-the-rest sort of technology. Users often had to do the rest themselves.

Of course, before people could experience the pride of ownership, the joys of the open road, or the agonies of car trouble, they had to buy an automobile. While a surprising number of individuals built their own horseless carriages, cobbling parts together in barns and sheds before venturing out noisily and awkwardly on local roads, the vast majority of early adopters satisfied their desire for motorized mobility by purchasing one already manufactured.[83] Then as now shopping for a car and buying one could be a fraught undertaking, not least because of its cost. With the exception of buying a house, consumers seldom shell out as much money for anything as they do for their automobile. Most significantly, early buyers were often quite baffled by the choices involved in selecting one. Choosing a particular type, make, and model from among the many versions of this recently introduced and highly complex technology, one about which they knew virtually nothing, was itself a user unfriendly experience. The challenges faced by car buyers, and the changing relationships they had with auto sellers during the early decades of the auto age, are the subject of the next chapter.

Chapter 2

Buying an Automobile

A S THE NINETEENTH CENTURY DREW TO AN END, people were getting excited about the next new thing: the horseless carriage. Unlike horses and humans, both of which had muscles, an automobile did not tire; if fed oil and gasoline, the endurance of this mechanical steed seemed limitless. Had not former bicycle maker Alexander Winton driven a horseless carriage all the way from Cleveland to New York in 1897 and then repeated the feat again in 1899? On that second trip, Winton took along a journalist who filed stories en route, spreading the word about the new technology and publicizing his newly launched Winton Motor Car Company. The horseless carriage was also fast, as Winton proved in 1902 by driving another of his machines, "Bullet Number 1," ten miles around a racetrack averaging over fifty-five miles an hour, nearly twice what horseflesh could do.[1] Nobody knew what the future of the automobile might be, but already in 1895, even before Winton's accomplishments, the founders of the first automotive trade journal in the United States were optimistic enough to title it *Horseless Age*.[2]

Many ridiculed such optimism and thought the automobile a passing fad, never likely to supplant the trusty horse.[3] Others were terrified by the noises emitted by internal combustion engines and thought it foolhardy to sit atop such explosions, a necessity given that the motors in early automobiles were usually directly under the seat of the operator and passenger. It also seemed unwise to sit in the open air while traveling speeds of twenty to thirty miles an hour without the protection of heavy steel all around one, as was the case in railway passenger cars.[4] Indeed, so dangerous did automobiles seem that some people wanted to ban their use altogether, such as the residents of Eden and Bar Harbor, Maine, who succeeded in doing so within their jurisdictions.[5] Many more municipalities and states passed laws restricting and regulating the use of the new technology, such as requiring cars to be licensed or establishing speed limits.[6]

Among those most hostile to the new technology were residents of the countryside. Many rural folks considered the automobile a moral menace or "devil wagon," while farmers resented the affluent city slickers who drove them, roaring over rural roads, terrifying livestock, and causing horses to rear and bolt and sometimes overturn buggies and injure or kill the occupants.[7] These rural constituencies were rumored to be responsible for the nails that occasionally turned up on country roads, puncturing the tires of automobilists and deepening tensions between city and countryside. In the cities themselves, residents of poorer, more congested neighborhoods, accustomed to congregating and playing in the streets in front of their tenements, fought the incursion of loud, fast-moving automobiles into what in effect were their front yards.[8] These conflicts sometimes turned violent, and the *New York Times* regularly reported statistics on "automobiles stoned."[9] Everywhere, in fact, the newfangled invention fostered tension and class antagonism. "Nothing," asserted Woodrow Wilson in 1906, then president of Princeton University and soon to be president of the United States, "has spread socialistic feeling in this country more than the use of automobiles."[10]

The doubt, skepticism, and hostility directed at the new invention did not, however, dampen people's desire to have automobiles. From the moment they were offered for sale, buyers eagerly stepped forward. While Europeans had invented the horseless carriage, it was Americans who per capita adopted motor vehicles more rapidly and in far greater numbers than did people anywhere else. Already by 1910 the U.S. automobile industry produced more cars than the rest of the world combined.[11] The first American firm to enter production was the Duryea Motor Company of Springfield, Massachusetts, founded in 1895. The brothers Charles E. and J. Frank Duryea were only two of the many Americans who had been experimenting with motorized buggies, but they were the first to actually produce automobiles for sale. The thirteen identical machines they sold in 1896 represented the first production run of any American automobile firm.[12] The industry expanded rapidly, however, and soon struggled to meet demand. By 1900, the first year for which we have statistics, consumers purchased forty-two hundred cars manufactured by thirty or so firms, including Duryea, Winton, Packard, Columbia, and Locomobile. Five years later the number of buyers had increased nearly sixfold, as some 24,250 men and women became new owners. By 1910 new car purchasers numbered almost 200,000, while in 1916 they for the first time exceeded a million. Consumer demand would continue to grow until the late 1920s, when the automobile market became saturated and sales slowed.[13] But by 1930 the "horseless age" had indisputably arrived: the majority of American households owned a car, those adults who did not yet have one dreamed of getting a car, and for better or worse the nation had embraced a transportation system based on privately owned motor vehicles operating on publicly subsidized highways, the earlier transportation system reliant on public trains and trolleys falling into desuetude and abandonment.

Early adopters bought motor vehicles for different reasons. Many relished the adventure, excitement, speed, and thrills of the open road and spoke of the freedom and power they felt when behind the wheel. Others had more prosaic motives. Automobiles enabled them to get around town more rapidly and conveniently than had been possible walking, with horse and buggy, or riding a streetcar. And unlike any public conveyance, a personal car was available on one's own schedule and was never crowded with strangers. With automobiles, people could also get out of town altogether, availing themselves of the pleasures of extended "touring" in the countryside. Businessmen and professionals touted the practical benefits of automobiles, for visiting clients, and physicians early discovered it to be ideal for calling on patients.[14]

But no matter what motivated people to acquire an automobile, their early involvement with the new and complex technology was invariably accompanied by vexation. Their struggles began as they shopped for their first automobile. During the years from the industry's beginnings around 1900 to about 1915, first-time buyers faced sometimes daunting challenges. One was the "absolute bewilderment with which the variety of choice oppresses" the neophyte buyer in his "selection of a motor-car," as one writer described this challenge in 1904.[15] This difficulty was compounded by the fact that automotive technology in this period was rapidly changing and the industry was groping toward some consensus as to what exactly the term *automobile* meant. For their part buyers had no way of evaluating untried and unproved designs. A 1905 Winton sales brochure asserted that the car had "no experimental features," but around the same time, one of the nation's earliest auto salesmen, Harry Shiland, felt the Buicks he was selling were still a bit too experimental: "You aren't selling cars to mechanics," he chided William Durant, the company's head, insisting that "cars have to be foolproof for the average doctor or lawyer or businessman to want them."[16] But they weren't, and so the dilemma of shopping for and buying one was great. "The purchase of an automobile is like breaking out of jail," as one man who endured the experience wrote, "a thing not to be lightly undertaken" and one that would likely have painful consequences.[17] Car buying from 1900 to 1915 was not only bewildering but, like breaking out of jail, terribly stressful.

People rarely consider their activities as consumers important enough to write about, but technology consumption often leaves a rich trail of evidence bearing on the experience. We can learn what types and makes of automobiles consumers chose even if their motives for making those choices remain hidden. We can read the same advertisements, promotional literature, and other books and articles they read in seeking advice as to what kind of car to buy, as well as the same "lessons from the road," as *Horseless Age* called articles written by motorists based on personal experience, intended to inform prospective purchasers and others who felt quite overwhelmed by the technical complexities of the new technology.[18] Shoppers also turned to published information like this in the 1900s in part because because actual horseless carriages were still rare and hard to see and compare.

"I commenced (as nearly everyone else does) by sending for catalogues of every machine that I heard of," one buyer commented in *Horseless Age* in 1904, but "looking all the catalogues over, I was about as much at sea as ever."[19] Like others confronting a complex new technology, he would have stumbled over the terminology used to describe and discuss an automobile. Although catalogs and advertisements were aimed at the neophyte motorists who comprised the early market, they never defined or explained technical terms adequately for beginners. Prospective buyers puzzled over references to a car's "exhaust valve," "crank case," "magneto idler," "coil," or "float-feed carburetor," all of which were alien terms signifying equally mysterious objects. Neophytes also had trouble evaluating a manufacturer's claims and assertions. Some were clearly puffery, like Winton's in a 1905 ad touting its car's "automatic 'Fool-proof' Motor" that "does its own work infallibly without 'tinkering' or adjusting."[20] Other statements might sound truthful but, like those in another Winton publication, raised as many questions as they answered. The redesigned clutch of the new Winton, the text explained, "will not require the frequent readjustments which are so annoying to drivers," and the "set screws which secure the clutch adjustment" have been "increased in diameter" so they no longer could "be broken off with an adjusting wrench" or "work loose and drop into the gears."[21] Prospective purchasers might be impressed by the manufacturer's honesty in acknowledging the problems of its earlier vehicles but had little way of knowing whether those defects had really been remedied.

Buyers would have found it even harder to evaluate automobile advertisements that masqueraded as editorial content in periodicals, a not uncommon practice in the early twentieth century. The 1905 Winton piece in *McClure's* was typical.[22] It had a title, author—one Loyd A. Thomas—and tightly packed double-column text that ran for a page and a half, just like a real article. The first sentence grabbed readers like a good news story: "'The Driver lost Control of his Car!' That's part of the Newspaper report on nearly every automobile accident. Doesn't it set one thinking?" The piece went on to note that drivers sometimes lost control because their automobile's controls were too complicated. Only then, once one's interest was piqued, did a reader come upon the statement "This is where the 'Winton of 1905' scores over all other Motor Cars"—serving clear notice that the piece was an advertisement. The Winton's speed "is controlled by Air-pressure," the pitch continued, and there were "no Gears to wear out, no Springs to weaken, no Levers to stick, at critical moments," language implying that other automobiles had such defects and were therefore less safe. "With this one Pedal alone," the Winton advertisement concluded, "and using the high-speed clutch you can run Four miles an hour, or Forty miles an hour, or any speed between these two." Due to the car's "simple" design, "a Youth could run a Winton the first time he rode in it, after an hour's coaching."[23] That the "article" was obviously an ad was visually confirmed at the bottom of the second page, where a large illustration featured the latest model of the company's production.

Novice readers could learn from ads like Winton's that there were differences in the way automakers designed and built a car's speed control, or throttle. But could consumers evaluate the alleged superiority of pneumatic speed control over other engineering solutions? How should they weigh similarly plausible claims about throttles, couched in similar technical language, made by competing manufacturers? There was no easy answer short of accepting the advice offered by one British author as to how to go about "selecting" a motor car. "Go to an expert whose character and knowledge are beyond question, and who does not receive commissions from any makers or sellers of motor cars," he suggested, and then "pay him a fee" and "let him advise you as to the make or type of car you ought to get." The author seemed to anticipate such advice would go unheeded, especially by American readers, who would "insist on being their own experts," but that result threw prospective shoppers back on their own confusion and left them struggling to comprehend the automobile's many complications.[24]

The challenge facing early car buyers was compounded by the fact that they could not turn to any published source of independent judgment about automobiles. Today, shoppers routinely find, online or in publications such as *Consumer Reports,* extensive tests of products along with recommendations of particular makes and models. In the first two decades of the twentieth century, buyers could read the short articles about new models of automobiles that appeared regularly in *Horseless Age* or *Motor,* but these were either totally descriptive, enumerating a car's technical features without passing any judgment, or written by the auto manufacturer's own publicists and filled with the same self-serving claims found in advertisements.[25] The only published source that compared different vehicles in the first decade of the century was the *Hand Book of Gasoline Automobiles,* an annual compilation put out by the Association of Licensed Automobile Manufacturers (ALAM), a trade group organized to maintain a patent monopoly.[26] The editors of the handbook acknowledged the many "questions that arise in the mind of the purchaser" and gathered facts and statistics about every vehicle produced by its members, presenting them so buyers could readily compare vehicles. Yet the information in the handbook, like that found in manufacturers' catalogs, was narrowly descriptive, technical, and quantitative. Buyers interested in a medium-sized touring car, say, might consult the 1905 volume and learn the price, seating capacity, weight, gasoline capacity, wheel base, tire dimensions front and rear, and type of frame, transmission, steering mechanism, and cooling and ignition system of vehicles fitting their needs, such as the Winton Type C and the Northern touring car. But once novices learned that the Northern employed "wheel and gear" steering while the Winton used a "screw and nut" design, or that the Northern's frame was fabricated of "angle steel" while the Winton's was constructed of "one-piece, channel section, pressed steel," what were they to make of such information?[27] Without considerable technical knowledge, how were consumers to use such information in their evaluations? As an engineer acknowledged in 1908, many of the "facts and figures" bandied about in marketing automobiles held little meaning for those without technical backgrounds.[28]

Neither the dry, technical facts and figures of the ALAM handbook nor the self-serving, exaggerated claims of manufacturers did much to dispel the bewilderment of car buyers. Articles and books written by motorists were probably the most helpful to novices when choosing a car, for while such literature almost never mentioned the make or model of automobile, it expressed an individual's point of view and was not commercially motivated.[29] Shoppers who picked up one such book, *The Complete Motorist*, found themselves presented with the useful question: "What is the car wanted for, and to what use is it proposed to [be] put?"[30] Novices who had never owned a car or even driven one might be unsure exactly how they would use a vehicle, but the question would enable, say, a physician who knew he would use a car to make house calls to drop large touring cars from consideration and look only at two-seat runabouts. By reading, novices could also learn that the issue of use in the first decade of the century was complicated by the fact that automobiles were more unlike each other than would later be the case. Some cars were hill climbers, others were not; some were suitable for touring in the countryside while others, like electric Broughams, were not, though they could elegantly and silently whisk their city owners to the opera or theater. Many automobiles of the time had insufficient ground clearance to manage high-crowned, unpaved rural roads, although so-called "highwheelers," light, buggy-like machines, did. On the other hand highwheelers and most electrics were slow, barely capable of reaching fifteen or twenty miles an hour, while steam and gasoline machines could easily hit forty. The widely varying capabilities of early vehicles not only circumscribed a car's potential usefulness but also made choosing more difficult for first-time buyers.

Prospective purchasers were also advised to focus their choice not on the automobile itself but rather on the integrity of the manufacturer, to "learn of the men who operate and direct" the firm.[31] Buy from a "manufacturer and not an assembler," one veteran motorist suggested—that is from a company that made most of the components of an automobile rather than one that bought engines, chassies, bodies, wheels, and other parts from others and simply assembled them, a common practice at the time. Additionally, he urged prospective purchasers to buy from a company that has invested enough in manufacture so buyers could be "reasonably sure that he will still be in the business when your car gets old and becomes worn, needs repairs and an occasional new part."[32] A similar desire to buy only from a solid firm animated a prospect in 1915 who was contemplating the purchase of a Dodge, a make that was not yet on the market and built by a new automaker. He promised the Dodge agent in Ann Arbor, Michigan, that he would buy the car "on the condition" that the "Dodge Brothers plant in Detroit is like the big picture" of the factory he had seen in the dealer's office. He probably did not know that for years Dodge had built bodies for Fords and was therefore hardly a start-up, yet his behavior was consistent with the belief that the vaster a manufacturer's grounds and more numerous its buildings, the less likely it was to go out of business and thus to "orphan" its cars and those who had purchased them.[33] This buyer may also have been heeding another piece of advice common in the early auto era, the "good argument in favor of buying an automobile that is made near one's home."[34] For him, an Ann Arbor resident, the Dodge factory in Detroit was just down the road.

Beginners were also advised, in the words of one expert, to start with a "two- or at most three-cylinder car of about 10 h.p." In the first decade of the century, fewer cylinders almost always correlated with lighter weight, lower horsepower, and easier handling and steering, all characteristics recommended for novices. Engines with two or three cylinders were also less mechanically complex than those with four, six, or more cylinders, and experts advised that a starter car be simple: "The simpler the machine and the less parts, the less the possibility of trouble."[35] First-time buyers were also advised to get a car in which, when trouble did occur, they had "the ability to remove any one part of the car without disturbing all others," a quality more likely with small, low-powered, vehicles. Buyers should never believe a salesman who says "his particular make of car never needs to be dissembled, that nothing wears out, that no accident can possibly happen," the expert added; to so believe was delusive, "a consummation devoutly to be wished."[36]

Early automobile buyers also puzzled over conflicting advice regarding two other design issues. The first was whether they should steer an automobile from the left side or whether the steering wheel should be on the right. Even though the rules of the road in the 1900s were no different than today's, requiring traffic to keep to the right and to pass to the left, most American-made automobiles in those years had their steering wheels on the right, like present-day British cars.[37] These included the Buick, Cadillac, Chalmers, Maxwell, Oldsmobile, Winton, and, before 1908, the Ford, although a few marques, or makes, such as Brush and Chandler, placed their steering wheels on the left, anticipating what would later become the American standard.[38]

"Each system has its advocates," asserted an article in *Automobile* summing up the arguments for what were called right-hand and left-hand steering in 1912, "and although the former has been considered standard practice for years, the latter has been gaining ground rapidly."[39] Yet why would anybody want right-hand drive on a car driven in the United States? For one thing, steering from the right-hand seat allowed operators to more easily gauge the distance from the wheels on that side to the drainage ditches that ran alongside the era's dirt roads. Given the crowned construction of those roads, a vehicle hugging the right side of the road was already disposed to drift toward the ditch, and if a wheel slipped into the ditch its wooden spokes would probably collapse, flipping the vehicle and perhaps injuring or killing its occupants. Operators of right-side-drive vehicles also felt more comfortable crossing narrow horse-and-buggy era bridges, which often lacked guardrails; on such spans drivers wanted to stay as far to the right as possible to allow oncoming vehicles to safely pass, but obviously they also wanted to avoid the potential disaster of their right wheels falling over the edge of the bridge. Finally, many early automobilists favored right-hand drive because they could more readily gauge the distance of their vehicle's right wheels to those of any slow-moving horse-drawn buggy or wagon they were overtaking and passing on a narrow road.

Advocates of left-hand drive pointed out how they could pull up to a curb and discharge or pick up passengers without getting out of their cars. With the steering wheel on the right, they noted, operators had two choices, each of them awkward or illegal. To avoid letting their passengers off to soil their shoes in streets fouled with horse dung, they had to pull up to the curb, climb out of the driver's seat onto the sidewalk, and have their passengers enter or exit the car through the right front door, maneuvering themselves to or from the left front seat. This was at best awkward and sometimes impossible, given that most automobiles had brake and gearshift levers protruding up from the floorboards in the center of the driver's compartment. The other option drivers of right-steer vehicles had in picking up or discharging passengers was to make a U-turn in the street and, proceeding against the flow of traffic, pull up to the curb so the left side of their car was parallel to it. Then a passenger could get out or enter the car directly from the sidewalk without disturbing the driver.

Two additional arguments supported left-hand drive. First, if gearshift levers were to be centrally mounted in the car, as increasingly they were, and because most people were right-handed, then shifting would be easier if operators sat and steered from the left side of the vehicle, enabling them to move the shift lever with their right hand. (Many early vehicles with right-hand steering, however, placed the shift lever, along with the emergency brake lever, to the *right* of the driver, on the edge of the chassis.)[40] Second, supporters of left-hand drive cited what in time proved the argument that clinched the debate: sitting in the left-hand seat while driving, it was much easier to avoid head-on collisions with oncoming vehicles.[41] Left-hand steering privileged the driver's view of the left side of his or her vehicle, the better to avoid such perils, while right-hand steering privileged the view of the right wheels, the better to avoid hazards to the right of the vehicle. As more and more automobiles were on the road, traveling upward of twenty or more miles an hour, head-on collisions at what could easily be a combined speed of fifty or sixty miles an hour became a far scarier prospect than ditches or oncoming buggies. City drivers touted a similar advantage of left steering in that it enabled them to better avoid collisions with heavy, rapidly moving trolley cars. Trolleys usually were double-tracked down the middle of the street in American cities, and motorists wishing to pass one had to pull out wide to the left around it to do so, risking collision with any trolley coming down the other track in the opposite direction. Before moving out from behind the trolley they were following, motorists obviously wanted to peek around it, something that was all but impossible if they had a right-hand drive vehicle. Indeed, urban and suburban motorists, along with dealers in such areas, appear to have most vehemently demanded the change from right to left steering. A Detroit dealer with considerable experience driving "in crowded streets in many of our American cities" believed "there should be a law to prevent people driving a car with right-side drive" as they "are endangering the lives of others."[42]

Along with the question of left versus right steering, car buyers puzzled over a second issue, what engineer-historian Gijs Mom dubs a "trilemma"—deciding whether to get an automobile powered by steam, electricity, or internal combustion.[43] Each of the three propulsion alternatives had ardent advocates among early automotive entrepreneurs, and early motoring magazines published articles by motorists recording their experiences, for better or worse, with electrics, steamers, and gasoline cars, as the three variants were commonly called. Although editorial policy often forbade authors from mentioning the name of automobile manufacturers in such articles, readers still could get a sense of the strengths and limitations of each general type of vehicle.[44]

Sometimes authors explicitly compared two types of vehicles, as did New Englander James Hamilton in his 1904 article in *Horseless Age* about his first two years as an automobilist. Although steam power had long been familiar in New England, driving mills, locomotives, and ships for decades, Hamilton initially had been "inclined toward a gasoline machine." But he changed his mind on hearing "that they were not as good on hills as steam machines." So he bought one steamer and then another before again changing his mind, having decided that the drawbacks of steam outweighed whatever advantages it gave on hills. "One of the things I disliked" about steam cars, he said, "was the everlasting firing up and getting ready." He was referring to the need to light the steam engine's boiler at least fifteen minutes before getting under way; if he owned one of the popular Stanley Steamers, he would first have had to light the pilot before lighting the main "burner" that heated water in the engine's boiler. This work was complicated by the fact the Stanley's pilot burner used gasoline as a fuel while the main burner used kerosene. Hamilton not only "tired of the work required in starting" a steam car but also complained of the need to keep "my eyes glued on the water glass, steam gauge, air gauge and many other things" while running the vehicle. He determined never to have "anything to do with a steam machine again."[45]

On the plus side, as many others pointed out, steam cars were silent. "As compared with a gasoline explosive engine," wrote one owner in *Horseless Age*, "the elasticity and quietness of steam render it far superior."[46] This so-called "elasticity," meaning steam's ability to keep working as it expanded, had implications for the ease of driving steamers compared to gasoline vehicles. "Mrs. J.C.C.," a California motorist, emphasized this point in another "Lessons of the Road" article when she compared the experience of running her husband's gasoline runabout with that of operating his steamer. The steamer was very complicated to start, she explained, but once started it was "easier to handle—no clutches, no speeder [throttle], no sparker [ignition control], etc., to think about."[47] One more advantage steamer enthusiasts often cited was that steam engines were paragons of simplicity compared to gasoline motors, having only a handful of moving parts. On the other hand, besides requiring an elaborate and complicated ritual to start, another "troublesome feature" of steamers, as a physician observed in 1910, was the need for "constant refilling of the water reservoir. . . . A trip of ten to fifteen miles would consume about one-half the contents," he claimed. For many motorists the steamer's profligacy in water consumption was an intolerable nuisance.[48]

We do not know how individual readers reacted to these accounts of the relative strengths and shortcomings of steamers, electrics, and gasoline cars, but we do know their collective response. During the first decade of the century approximately one out of ten car buyers, about twenty thousand people, chose steam. The vast majority of them bought vehicles produced by the three leading manufacturers: Stanley (fifty-two hundred), White (over nine thousand), and Locomobile (over four thousand). The remaining buyers of steamers got machines built by Mobile, Lane, Victor, and literally dozens if not hundreds of smaller firms, some little more than machine shops that produced a few cars hoping to establish a toehold in the burgeoning automobile business.[49] Steamer owners, therefore, were especially likely to wake up one day and discover that the manufacturer of their machines had gone out of business and their automobiles were thus orphans, after which parts would become difficult if not impossible to find. When the largest maker of steamers, White, felt the tide turning and in 1911 shifted exclusively to producing gasoline-powered vehicles and trucks, it created another large group of orphans. Stanley continued to build handfuls of steam cars until 1925, but for all intents and purposes by the early 1910s the trilemma of choice once faced by car buyers had become a dilemma: gasoline or electric power.

From 1900 to 1915, about twenty thousand consumers bought electrics, about the same number that purchased steamers. In terms of market share, however, electrics early on had been more popular, making up about 5 percent of all new car sales in 1905. Thereafter, the market share of electrics declined steadily, and by 1912 only one percent of car buyers opted for electric power. In terms of total sales, however, more people bought electrics in 1912 and 1913 than ever before—about six thousand each year—yet in those same years over 90 percent chose gasoline-powered cars.[50]

The appeal of electrics had been the greatest in the early 1900s, when they compared favorably with steamers and especially gasoline cars. Electrics were as quiet as steamers and did not belch noxious fumes like gasoline cars. They also were far simpler to run than the other two alternatives. The ease and alacrity with which one could get under way was almost miraculous—operators just flipped a switch and drove off, without having to crank an engine or wait to raise a head of steam. Once under way, operators manipulated but a single lever, essentially a rheostat, to control the electric's speed, and as with the steamer there were no clutches or gears to shift. One steered early electrics with a tiller, a device perceived to be easier and more intuitive to use than a wheel. Furthermore, electrics were enclosed and thus offered protection from the elements and weather, unlike mostly open steam and gasoline cars. Additionally, as a physician wrote of the electric in 1910, "It is much easier to look after. You do not have to look after oil; you do not have to watch the gasoline and you have no spark trouble. I run the car into the barn, shove the plug in and it fills itself full." Indeed, he enthusiastically concluded, "For a doctor an electric car is the only car."[51]

He did not mention the hazardous work of having to check the level of electrolyte in the electric's batteries and when necessary adding distilled water or acid. This and other disadvantages were little discussed in popular periodicals or even in the automotive journals. Charging an electric obviously required access to the electric grid, so consumers living on farms or in small towns without electrical service—still a majority of Americans in the first decade of the century—could not consider owning such a vehicle. Although electrics were relatively trouble-free, repairing them required specialized equipment and knowledge. Electrics were also relatively slow compared to the other two types, although their predominantly urban buyers did not seem to have found this a problem. The range of electrics, forty or so miles between charges, while subsequently often cited as the major reason for their lack of wide acceptance, was also not much discussed. In fact, electrics could go further between charges than steamers could between refills of water, though it was often easier to procure water (or even gasoline) on the road than to find a place to recharge batteries.[52]

What, then, prompted the overwhelmingly majority of buyers in the first two decades of the century to choose the third option: cars powered by internal combustion? The gasoline car's advantages—speed and range and, most notably, its being generally less expensive—had to be balanced against many negative qualities: gasoline cars were noisy, rough-running, polluting, more complicated, and much more difficult to operate than electrics or steamers. To start a gasoline car, operators had to crank the engine, a potentially dangerous and often arduous task. And once under way, operators continuously had to shift gears. There is therefore no easy answer as to why consumers chose such vehicles over their steam and electric rivals. The "technology choice," as sociologists call a collective decision like this in favor of internal combustion, did not result from any inherent technical superiority that was recognized by the market. Notions of "best" or "superior" are not intrinsic qualities of technological artifacts; they are social constructs, the result of people's values, culture, and traditions, all of which played a part in favoring automobiles powered by internal combustion.[53]

One of the factors shaping the preference for gasoline cars, as historian Virginia Scharff has persuasively demonstrated, was people's attitudes about gender. She recounts the almost universal belief on the part of men, especially those in the automobile industry, that electrics were for women. The fact that electrics were so strongly gendered female meant that men, who still comprised the majority of car buyers, tended to prefer gasoline-powered automobiles. Advertisements for electrics invariably depicted elegant, well-attired ladies as owner-users. Henry Ford, who personified the internal combustion–powered automobile as much as anybody, bought an electric for his wife, Clara, as did his friend, Thomas Edison, for his wife. Men believed the limited physical exertion needed to operate an electric, along with its limited range and speed, best fit women's place and role in society. Quoting C. H. Claudy, the automotive columnist for *Woman's Home Companion,* who claimed, "It would not be amiss to call the electric the modern baby carriage," Scharff shows how men viewed motoring women as threatening and tried, in effect, to infantilize them by confining them to vehicles that restricted their movements. At a moment when women were pressing for the vote and equal rights, however, not a few female car buyers rejected the slower-moving, more ponderous electric. Some statistics suggest female buyers were even less likely than men to choose one, and some purchased larger, high-powered gasoline cars as a kind of protest, wanting nothing to do with a vehicle that could be considered a "baby carriage" for women.[54]

Historian Mom suggests another reason men preferred gasoline cars. A "male sports culture of technical challenge and adventure" existed in the early 1900s, he believes, one that relished speed and racing and rewarded the solution of technical problems. Such a culture, he further argues, was well adapted to the complex and cranky gasoline automobile and encouraged male buyers to minimize the drawbacks of gasoline power and ignore what he calls the "engineering rationality" of the electric automobile.[55] Even if their choices were technically irrational, consumers by the hundreds of thousands had by the early 1910s created an all but unstoppable momentum in favor of internal combustion. In 1912, when Cadillac introduced electric starters on its gasoline cars, co-opting the most attractive feature of electrics, which was their ability to instantly get under way, the issue of technology choice was resolved. Employing a storage battery and electric motor, electric starters eliminated the work of hand cranking internal combustion engines. Beginning with other General Motors companies, automakers quickly followed Cadillac's lead and made electric starting standard equipment. But it was not until the latter half of the 1920s that the majority of new cars started at the turn of a switch or with the press of a button. The spread of electric starters paralleled a precipitous slide in electric-vehicle production, and already by the mid-1910s sales of electrics fell off dramatically, and two of the three leading firms in the field had ceased production.[56]

By about 1915, then, prospective buyers no longer were likely to puzzle over the power source, as the concept of an automobile in the United States had become stabilized as a moderately large vehicle, powered by an internal combustion motor running on gasoline. Such a car would have at least four cylinders and considerable power compared to English or European vehicles. Even the inexpensive Ford Model T had twenty-five horsepower, which, in the car's lightweight chassis and body, provided peppy performance. Beloved by beginners as well as veteran motorists, the Model T helped eliminate those one- and two-cylinder starter cars once recommended for beginners. With its high road clearance and clever three-point suspension, the Model T could also handle the worst rural roads, putting firms that had sold high-wheel, buggy-type automobiles out of business. By 1915 the difference between a "country" or "city" car had ceased to exist, as like the Model T virtually all automobiles were suited to a range of purposes and environments.

In numerous other ways, the Ford Motor Company and its Model T helped bring the early period of car-buyer confusion to an end. Initially introduced in 1908 and mass-produced on moving assembly lines starting in 1913–14, the Model T sold some 15 million units over its production life to 1927. The earliest models sold for a hefty $825, but Henry Ford introduced efficiencies in manufacturing, plowed the savings back into his business, and each year lowered the sales price of a new car. In 1922 one could buy a new roadster for just $290, the equivalent of only five months' wages for a blue-collar worker.[57] Ford's Tin Lizzie literally put the nation on wheels, as from 1913 through the early 1920s more than half of the American auto industry's *total* production consisted of Model Ts. By selling so many Model Ts, Ford defined basic, low-priced American automotive transportation.[58]

The Model T not only established the four-cylinder gasoline engine as the norm for low-priced cars but also closed the debate over right- versus left-hand drive. Henry Ford originally built vehicles with right-hand steering but in 1908 had switched to left-hand drive with the debut of the Model T. As sales of Ts soared in the early 1910s, the percentage of cars with left-hand steering rose, and most manufacturers rapidly fell into line behind Ford and switched to left-hand drive, although as late as 1912 at least 110 American manufacturers retained right-hand steering.[59] In 1913 Buick, Winton, and Packard adopted "the convenient left drive," as a Packard ad called it, while Cadillac and Locomobile did so in 1915.[60] "Left Drive with Right Hand Control is used—the best liked arrangement," explained a Locomobile ad from that year: "The gear levers are operated by the right hand, which is safest and most natural."[61] By the latter part of the decade, only expensive marques like Pierce-Arrow stuck with right-hand drive, either as standard or as an option, perhaps to accommodate their wealthy clients, who often shipped their cars to England for touring.[62]

Ford's Tin Lizzie also helped establish a baseline for standard equipment even though many people derided its lack of a speedometer, gas gauge, electric starter, and other accessories. Yet the Ford did come with many essentials that had not been common in the early years of the industry, when automobiles routinely were sold without windshields, horns, lights, cloth tops (if open vehicles), speedometers, and even wheels, a situation that confused novice buyers who were unsure of their needs and what to buy. The purchaser of a new 1905 Winton that came without a canvas top or protective side curtains, for instance, decided after being caught out in the rain a few times that he needed such items and wrote the factory to order them. He received a reply from Winton, saying, "We do not manufacture tops" but passing along the name of a few firms that might be able to provide one.[63] Whether he eventually procured a top for his Winton isn't clear, but his experience underlined the total absence in the early auto industry of any agreement as to what comprised essential, or standard, equipment. The Winton Company itself seemed confused on the matter, having promoted its 1904 touring car saying it "will include, as part of the regular equipment, the finest detachable canopy top possible to make" because "a touring car should be for all weathers—rain or shine" and then concluding with the rhetorical question: "without a covering, with side curtain attachments, etc., . . . can any automobile be properly called a touring car?"[64] Winton answered emphatically in the negative, yet the very next year it advertised its 1905 model as available with or without a top! Model Ts, whether roadsters or touring cars, all came with canvas folding tops as standard equipment; as a 1913 Ford advertisement put it, "No Ford Cars Sold Unequipped."[65] But this meant only the equipment Henry Ford believed to be essential, and his personal tastes were Spartan. All Fords came from the factory with lights, wheels, windshield, and canvas top if an open car but not with detachable rims, electric starter, or speedometer, the latter instrument being one Henry thought an unnecessary extravagance. Belatedly, in 1919 he made electric starting standard, but only on closed coupes and sedans, not roadsters and touring cars. Finally, the feature became standard on all Model Ts in

1926, a year before the car went out of production. Nevertheless, Ford's share in the market and the fact that other makers followed his lead meant that, already by the middle of the 1910s, car buyers no longer had to worry whether their new automobile came with the basic essentials.[66] The period in which car buyers were bewildered by the many technical choices they had to make about design and equipment had come to a close.[67]

Yet in 1915 the "Horseless Age" so optimistically predicted twenty years earlier had not arrived. Some 2.3 million cars were registered in the United States that year, but many more adults still walked or traveled by horse and buggy, bicycle, or trolley car or train than owned and drove automobiles. The public was enthusiastic about automobiles, however, and motoring had become a common topic of conversation. "Of the three languages," the Ford Motor Company rhetorically asked its car owners: "do you speak—English—Baseball—or Motor?" Obviously they spoke and understood "Motor," as did millions of other Americans.[68] Whereas earlier in the century only automobilists knew terms like *spark plug* and *carburetor* or phrases such as "starting on the spark" and "spinning one's wheels," the vocabulary of motoring could now be heard almost everywhere. The language of "Motor" also embraced the proper names— "Ford," "Cadillac," "Packard," "Studebaker," "Winton," "Pierce Arrow," and so forth—that designated different marques. An automobile usually carried its brand name visibly on its body, sculpted in metal letters or emblazoned on a painted escutcheon or nameplate. It advertised itself and enabled any curious observer to identify its make. Many people learned to recognize an automobile's make even at a distance, noting the shape, size, or location of its radiator, headlights, fenders, or windows, and the general form of the body.[69] "There's the new Maxwell or Hupmobile," they could say, as traffic rolled by.

Speakers of "Motor" also used automobile brand names as shorthand labels for a person's class or social rank, for Americans assumed people were what they drove. To say of somebody that he drove a Ford was to position the owner of that no-frills vehicle in the lower part of the socioeconomic ladder, while noting of somebody that she drove a Pierce-Arrow, Peerless, or Packard placed her near the pinnacle of the class hierarchy. A popular 1913 song, "The Packard and the Ford," relied on such shorthand, relating a romance between the two cars at opposite ends of the cost spectrum and social hierarchy. Although the gimmick of the tune is silly, the "male" Packard falling for the "female" Ford, it echoed plausibly with a public that recognized that successful men often married down, choosing brides from less affluent families.[70] This Packard-Ford juxtaposition was frequently invoked before World War II to signify the two extremes of class, as when a Model T owner complained in a letter to the factory but made it clear that "I don't expect Packard luxuriance for Ford prices."[71] Other automobile names may not have designated class with the sharp specificity of "Ford" or "Packard," but they still resonated in people's imagination and showed up in popular songs. Some of these tunes, like "In My Merry Oldsmobile," from 1905, became hits, while others such as "In a Hupmobile for Two," "Mack's Swell Car Was a Maxwell," "Take Me Out in a Velie Car," "Over the Overland Route in an Overland Car," or "My Studebaker Girl" reminded listeners of particular brands, augmented their fluency in "Motor," and celebrated the status conferred by owning a car.[72] While most people probably knew, as another song title put it, that "You Cannot Judge a Man's Bank Book By His Au-To-Mo-Bile," doing so was already an American habit.[73]

During the 1920s, automobility became a truly mass phenomenon, the "Horseless Age" having finally arrived. The number of car buyers soared and automobile registrations increased seven times faster than the population. In every year after 1922, consumers purchased over 2 million new vehicles. Whereas in 1915 only a quarter of households owned a car, by 1930 fully three-quarters did and the "family car" had become a national institution.[74] As people increasingly got behind the wheel of their own vehicles, they remade the landscape, starting with the residential neighborhoods in which they lived, building garages adjacent to or behind their homes in which they housed their beloved machines.[75] They voted tax dollars to pave old roads and to build new ones; they even tolerated the erection of signs and stoplights to regulate themselves as part of ever-growing streams of traffic.[76] Their collective behavior behind the wheel also gave rise to new, auto-centric commercial environments, such as the shopping plazas with abundant off-street parking that were emerging on the fringes of old shopping districts and outside of downtown areas.[77] And not least, during the 1915–30 period these millions of consumers— most of whom were driving the first cars they had ever bought—participated in the dramatic reshaping of the practices and rituals of car buying into their present-day forms.

To understand those changes, we need to emphasize how different car buying was at the outset of the auto age in the early 1900s. Then buyers purchased vehicles in a greater variety of ways and through more channels. Some buyers purchased automobiles from salesmen who sold them door-to-door, just as peddlers had sold sewing machines and clocks. A few bought cars from department stores such as Wanamaker's, in Philadelphia, although the company abandoned the practice in 1904 due to the complexities posed by having to perform after-sale maintenance and repair.[78] Others bought their automobiles directly from the factory, as early auto manufacturers sometimes sold both retail and wholesale. The majority of very early buyers, according to automotive historian James Flink, however, purchased cars at automobile shows. Trade shows featuring a single product had been pioneered by bicycle makers in the 1890s, and the automobile industry seamlessly adopted the practice starting in 1900 with a show in New York City's Madison Square Garden. At least five other automobile shows occurred that year, "all advertised as 'firsts,'" Flink wryly notes. Because the public's demand for machines in those early years outstripped supply, it was a seller's market, and automakers used shows to take orders rather than engage in systematic marketing. At the 1903 New York auto show, a start-up that had not yet sold a single automobile and had only a single prototype took thousands of orders along with cash deposits from prospects, but this provided capital enough for the recently founded Cadillac Automobile Company to get into production, and the rest, as they say, is history.[79] Yet already by 1905, according to Flink, the majority of car buyers were buying from local manufacturers' "agents," or "dealers," terms that were used interchangeably at the time.[80]

Those early dealers rarely resembled what today we think of as car dealerships. Most were not devoted exclusively to the selling and servicing of automobiles, and few had the facilities that soon would come to be associated with such businesses: a showroom to display new vehicles, a garage to maintain and repair the cars sold, a waiting room for customers, or a parts department. More characteristic of the pre-1915 years were businesses like that of "Dr. Charles F. Steele, Dealer in Automobiles and Bicycles, fine repairing a specialty," as the Alpena, Michigan, resident's letterhead described him. Steele may have had a workshop but does not appear to have had a garage, yet in 1905 he was an agent for the Winton automobile, built in Cleveland. He did business by buying a single car wholesale and then trying to sell it at the manufacturer's advertised list price, the difference going for his overhead and profit.[81] A few early agents were not really in business at all but simply had bought a car for personal use and made an arrangement with the manufacturer to get a commission if they facilitated any subsequent sales.[82]

Like Steele, however, typical automobile sales agents or dealers already had another livelihood. Many were merchants, a category of businessmen who traditionally added and dropped lines of goods as consumer interests shifted. This was the case with Staebler & Sons, of Ann Arbor, an early Michigan dealer. The senior Staebler founded the firm in the 1880s to sell and deliver coal in the small university town, and in the 1890s he took on the selling and servicing of bicycles. Noting the growing popularity of boating on local lakes, he then added small gasoline marine engines and boats to his line of goods for sale. Around 1901, his son bought a Toledo steam car, which led to the firm's selling of motorcycles and automobiles.[83] Staebler & Sons sold the Toledo until 1906, when the operation became an agent for Reo, a car built in nearby Detroit. "We want the very best line of automobiles we can get," Staebler senior wrote in a letter to the United States Motor Company in 1911, looking for new opportunities, and "would be pleased to have you inform us which of your makes of cars this territory is open for."[84] After selling a variety of automobiles, including Apperson, Matheson, Warren, Whiting, and the Waverly Electric, Staebler & Sons split off its coal business in 1915 and thereafter settled into a more or less monogamous relationship selling the products of the new Dodge Motor Car Company.[85]

Consumers who bought cars from Staebler & Sons or similar dealers also selling coal, bicycles, boats, animal feed, or tractors have left us little evidence as to what it was like to shop in such general store–like environments. Buyers may have found such firms familiar and neighborly, yet such businesses were often too rooted in the horse age to effectively support the sales and servicing of the complex machines of the auto age. Some of the most old-fashioned dealers paradoxically sold Fords, simply because the Ford Motor Company had established sales agencies at almost every crossroads and in every hamlet in the country. In 1910 Ford already had over a thousand dealerships, and by 1920 it had increased the number of outlets for its vehicles to more than six thousand; the number peaked five years later at about ninety-seven hundred, at which point half of all the automobile retailers in the nation sold Fords.[86] Many of these Ford agencies, however, were like the one in Glasgow, Kentucky, a tobacco-farming town of some twenty-three hundred inhabitants. "This dealer is not a live wire," wrote the Ford Motor Company's "roadman" from the Louisville office who regularly visited every agency and reported back to Ford headquarters. The "firm is interested in the sale of buggies, wagons, and hardware, not trucks, tractors, or autos," he explained, recommending that Ford secure new representation in the community.[87] The Ford Motor Company tried hard to winnow out those who were not live wires and to compel its dealers to sell only Fords and not other lines of merchandise, let alone other makes of automobiles. But in small-town rural America many Ford dealers continued to function into the 1920s as general stores, selling whatever might make a profit.[88]

Car buyers benefited from the proliferation of dealerships and from the fact that so many sold multiple brands of vehicles. By the 1920s, almost everywhere people had a new-car retailer close at hand, usually a Ford dealer. A small lumber town like Alpena, Michigan, on the shore of Lake Huron, that in 1905 had but a single automobile dealer—the Mr. Steele who sold Wintons along with bicycles—by 1907 also had a Ford agency and, by 1927, half a dozen firms representing the Buick, Chevrolet, Chrysler, Dodge, Ford, Oakland-Pontiac, Oldsmobile, Overland, Reo, Studebaker, and Willys-Knight automobiles as well.[89] Similarly, in the small metropolitan area of Champaign-Urbana, Illinois, the single automobile seller of 1906 had by 1921 been augmented by no fewer than nineteen new car dealers.[90] Even in poorer parts of the country car buyers had more choice. In the county seat of Oxford, Mississippi, in the 1920s, dealers handled Ford, Lincoln, Dodge, Whippet, and Chevrolet automobiles.[91] To be sure, it was only in the larger cities such as San Francisco, which in 1921 had sixty dealers stretched along its "Auto Row," that buyers might hope to see examples of a majority of the makes (108 in 1920) manufactured by the American auto industry.[92] Everywhere during the decade, however, prospective car buyers found it easier to see and compare cars without traveling to the annual auto show, usually held in the state or region's largest city, than it had been ten or fifteen years earlier.

AT THE SAME TIME that prospective buyers were gaining increasing access to automobiles, the experience of car buying was changing as a result of the numbers of vehicles being sold and the millions who were getting behind the wheel for the first time. Increasingly, most people who could afford a car already owned one, so few buyers any longer were "virgins," as salesmen crudely called those who had never owned an automobile or learned to drive.[93] In short, the market was saturated, forcing automobile manufacturers and car salesmen to work much harder to sell new vehicles. As a result buyers faced more aggressive sales tactics when shopping for cars. Indeed, the stressful, unpleasant aspects of dickering with car sales personnel long associated with buying a car begin in the 1920s.

Management utilized different strategies to try to control the changing automobile market and sell cars to newly "motor-wise car owners."[94] One strategy was to better train dealers and their salesmen in the psychology and practices of selling by publishing sales manuals. This was not a new approach, for companies like Singer Sewing Machine and the National Cash Register Company had been using sales manuals since the late nineteenth century to aid in selling complex new machines.[95] Automobile manufacturers had always provided manuals for *buyers,* but in the more competitive market conditions of the 1920s they, too, increasingly issued manuals to sales personnel as well.[96] These sales manuals detailed the specifications of a company's products, provided talking points for its salesmen, and revealed how automakers sought to remake their dealerships into environments where salesmen controlled every aspect of their interaction with prospects with the goal of getting them to sign a sales contract. Salesmen were asked to treat the "subject matter" of these sales manuals "as confidential." The publications were theirs only "while engaged in the sale of Oldsmobile Passenger Cars," as one manufacturer typically stipulated.[97] The texts offered a template as to what salesmen should say and how to act when a prospect approached the building: "Open the door to callers," a Chalmers manual instructed; be a "business host" to your visitors, implored a Ford manual.[98]

Reminding the often youthful, invariably male sales force to be polite was good for customer relations, but the primary purpose of manuals was to provide salesmen with what was referred to as "sales ammunition." Car buyers were depicted as the enemy whose resistance to buying had to be vigorously attacked and defeated. There is "ammunition in this issue," noted one in a series of Dodge sales publications, "used and approved by the front line men—men who have gone over the top" like infantry in the recently concluded world war. If the "ammunition" is not "the right caliber to fit your particular gun," the text exhorted salesmen, "All right, just change it so that it *will fit*—change it to fit *your* personality and *your* problems—and then—*use it*. Ammunition in the caisson does not win the engagement. Load it—aim it—and *fire*."[99] These military metaphors sought to bring soldierly drive and ferocity to the work of selling automobiles, work that customarily was broken down into three components: presenting the sales talk or "showroom demonstration" of the vehicle, giving a "demonstration ride," and closing the deal. Manuals usually provided fictionalized scripts portraying a salesperson interacting with a prospect for each of these three stages. Salesmen were not meant to memorize them but to follow the rhetoric and mannerisms demonstrated in the scripts for the effective sale of automobiles.

Manuals provided facts and arguments to help salesmen point out a vehicle's advantages and rebut claims that a competitor's vehicle might be superior. Customarily, manuals compared in detail two or more automobiles made by competitors that were similarly priced. The *Graham-Paige Sales Manual*, for example, compared the car with the Auburn, Pontiac, and Oldsmobile, vehicles prospects might be considering along with the Graham-Paige.[100] In a climate where automotive technology was in flux, sales manuals provided ammunition to refute buyer suggestions that a company's products might be out or date or otherwise inferior. In 1925 Dodge salesmen were encountering shoppers who believed that their car's competitors, the Nash and Overland, were superior because they had six cylinders while the Dodge had but four. Salesmen were advised to point out that the number of cylinders had "little if any bearing on *ultimate* satisfaction" and that a six-cylinder engine did not necessarily provide any more power, greater acceleration, or smoothness of operation than a four.[101] Model T Ford salesmen also were given a litany of responses to refute shopper criticisms of the car that, while revolutionary and exciting when it debuted in 1908, by the early 1920s had changed little and was being compared unfavorably to the Dodge and Chevrolet.

When asked why there was no speedometer, for instance, Ford salesmen were instructed to say speedometers were "not used as regular equipment on Ford cars because hundreds of thousands of owners do not use them," and that the instruments "are often out of repair" anyway. To queries as to why Fords lacked electric starters, dashboard lights, water pumps, and other items found on the cars of their competitors, salesmen were urged to point out that if those items were offered as "regular equipment, the prices of all Fords would have to go up." Buyers should also be reminded that "there are mansions and palaces and there are *homes*," as the sales manual put it, and most car buyers' "pocketbooks are in the home class."[102]

Along with providing lines for salesmen to speak as if actors in a play, manuals frequently also offered what in effect were stage directions as to where to stand and what gestures to employ. "Stepping toward the car but stopping at a point ten or twelve feet from it—about opposite the headlight, but a little in front of the car," read a passage in the 1916 Chalmers manual. "This position is important and should be carefully chosen." This was the spot from which prospects could best appreciate the "smoothly flowing lines" of the Chalmers and observe how those lines were in "harmony with what is called today the 'stream-line' effect," according to the fictional salesman in the script.[103] The buyer's interest in styling would only increase in the 1920s, as Detroit began to alter the external appearance of cars more frequently and move toward the annual model changes that would be institutionalized by the early 1930s. In the fictional drama between salesman and prospect in another manual, *Selling Chevrolets,* from 1926, the car buyer gets maneuvered to the same, front-quartering position about ten feet away from the car, a spot that was becoming an iconic vantage point from which to observe or photograph a car. "Mr. Prospect," the fictional Chevrolet salesman says, "I want you to stop here a minute and observe the beautiful streamline design and the high hood. The Chevrolet is a good looking car." It was also a sedan, like most cars by the latter twenties a closed car, so interior design joined admiration for exterior lines in sales rhetoric. "Won't you get into the rear seat, Mr. Smith?" After Mr. Smith steps into the rear seat of the car, the salesman says, "You notice how easy it is to close the door with that pull-to handle."[104]

Although car buyers may have been newly focused on interior comfort and external style, some still were interested in the technical aspects of automobiles. Showrooms in the 1920s often displayed a bare chassis, that is, a car with its body removed, which enabled customers to closely examine its engine, brakes, and running gear.[105] "It is not necessary to go into mechanical features with every prospect," *Selling Chevrolets* noted, "and with many it will injure the sale to do so." As early as 1912, a sales executive of the National Motor Vehicle Company asserted automobiles could be sold "without lifting the hood." Just as buyers of watches had no need to know "what kind of a spring" or "how many gear wheels" a given timekeeper had, car buyers had no need to know anything about a vehicle's mechanism, he believed. At the time his industry peers considered the idea "radical," yet it was prescient; within a few years most car buyers would no longer want the "chassis demonstration," as Chevrolet called it, and would be uninterested in what was under the hood.[106] Not surprisingly displays of a bare chassis slowly became less common in auto showrooms.

Even though they cared less about what was under the hood, car buyers always had been interested in a demonstration ride, but this sales ritual also changed significantly over time. Up until the 1920s these were literally "demonstration" rides as prospective buyers were more likely not to know how to drive. They therefore could easily fall prey to what an expert in 1905 called the "wiles of the demonstrator." Unscrupulous dealers changed the gears in a car's rear axle to impress buyers looking for a good "hill climber," and only after purchasing the vehicle (at which point the dealer would give him a few instructions how to drive) would the buyer discover that the machine was hopelessly slow on level ground because of the lower gears that had been installed to sell it.[107] Early buyers also were misled by the so-called "flexibility test" intended to show how slowly an automobile could run in high gear, something we'll have more to say about in the next chapter. "It is the easiest kind of a thing for a foxy driver to fool the inexpert prospect," explained the famous auto racer Barney Oldfield, "and to give him a demonstration of speed three miles an hour on high, when in reality he has slipped the clutch on the novice."[108] By the 1920s motor-wise shoppers were less likely to be fooled by such deceptions. Most important, although sales manuals continued to refer to the "demonstration ride" or "road demonstration," prospective buyers had begun to insist on taking the wheel themselves to feel how well the car performed. Recognizing this new reality, a Chrysler manual advised salesmen to "plan your demonstrations carefully" and to "drive home the point that you are going to take him for a demonstration ride—that it's necessary." Chrysler laid out eleven steps for the demonstration, carefully sequenced so that the salesman could show off the car's performance on flat and hilly terrain as well as on smooth and bumpy surfaces and, if possible, on a "slippery" or even "icy" street. Only at the final, eleventh stage of the scripted demonstration does the manual say to let the prospect take the wheel.[109]

From the car dealer's perspective, the showroom demonstration and demonstration ride were but means to the end of getting the prospect to sign a sales contract. No matter how much shoppers admired the lines, ride, or any other aspect of an automobile, until they signed a contract there was no sale. In the newly competitive automobile marketplace, the industry vigorously sought new ways to close deals and get buyers to sign contracts. One method was a "special room for closing contracts," like that built into the new sales facility of a St. Louis dealer. Hailed in the automotive trade press, the "isolation room," as it was called, was a windowless space where there was "nothing to distract the prospect," thereby enabling salesmen and managers to put maximum psychological pressure on buyers to get them to sign.[110] Sales manuals also were part of the new method and pushed new tactics for closing. *Selling Chevrolets* urged salesmen to "Ask for the order—often," in the words of a caption under a photograph showing a salesman proffering a pen and unsigned contract to a prospect. The manual contained three such pictures of a salesman thrusting a contract in the face of a beleaguered buyer. "Ask for the order immediately after the road demonstration, when your prospect's appreciation of the car is at its highest," read one caption.[111]

Manuals also suggested tricks to get hesitant prospects to sign. "Another way to close, where you seem to have the prospect ready but he won't sign," a Ford manual explained, was to get the potential buyer to answer some "minor question which means the greater answer." In the manual's example, a Ford salesman visits "Mr. Blank" at his place of business and asks, "Mr. Blank, may I use your telephone please?" The script with stage directions continued:

(Call your dealer on the phone.) "Hello, Mr. Smith, I'm at Mr. Blank's . . . . No, I haven't secured his order but I want to know if I can get him a Tudor with green wire wheels (or any other accessory in which the prospect seems especially interested) this week." (To Mr. Blank) "Do you prefer green or red wheels, Mr. Blank?" If he states his preference he has practically given you the answer to the major question, that he is ready to buy. At that point, the salesman can once again present your order blank with every degree of confidence that the prospect is buying of his own good judgment and merely show him where to sign.[112]

The "Double Close," whereby the salesman brings in the "sales manager or proprietor" to help pressure reluctant buyers, was another favored trick. "Very often," Chrysler salesmen were told, "the mere presence of one of the higher officials of your organization has a reassuring effect on the prospect. He likes to do business with the 'boss,'" as it makes him "feel more important."[113]

Whether car buyers felt reassured and more important by such double-teaming or just badgered is debatable, but closing rooms and the pressure tactics developed by the auto industry (which, of course, continue today) made car shopping by the 1920s a trying, even painful experience. It also was more complicated because of two new factors. The first was financing. Consumer credit expanded tremendously in the decade, and by 1925, 70 percent of furniture, 75 percent of radios, 80 percent of phonographs, 80 percent of appliances, and 90 percent of pianos were bought on time, along with nearly 75 percent of all automobiles, new or used.[114] Car buyers in the first decade of the century, it will be recalled, had to pay the entire list price up front in cash, which limited ownership to successful business and professional people or those who were independently wealthy. The Model T's success provided the catalyst to change this. Although Henry Ford himself despised bankers and insisted on selling Model Ts strictly for cash, some of his higher-priced competitors in the 1910s—the makers of Studebaker, Chalmers, Maxwell, Overland, and Paige—began to sell their cars on credit so as not to lose even more sales to Ford.[115] In 1919, General Motors jumped on the credit bandwagon, forming the General Motors Acceptance Corporation (GMAC) to finance sales of its cars. Soon even Henry Ford, for all his distaste of banks and credit, recognized the benefits credit offered in selling cars to people who could not pay all cash up front but who would be happy to buy if they could pay over time.

Ford first experimented unsuccessfully with layaway, a plan whereby buyers put down five dollars a week until they had paid the full price, at which point they got their car. "Our customers would accumulate $25 or $30," a Stockton Ford dealer reported, "then come in with a hard luck story to get the money back," leaving the vehicle unsold.[116] By 1925, Ford had abandoned his opposition to the new dispensation and quickly the majority of Model Ts were sold on credit provided by the firm's "Deferred Payment Plan."[117] Buyers of used cars also routinely obtained credit during the 1920s; indeed, they were often pushed into taking out loans. In a Kansas City firm managed by a "young woman who knows how to sell used cars," one Mrs. W. J. Birrell, "less than one in ten prospects get away without buying" because she was willing to take a personal note from buyers who, although they qualified for "a first mortgage" on the used vehicle, couldn't come up with the "remaining $25" down payment.[118] That both new and used cars were now sold on credit testified to the fact that automobiles had become sturdy enough—true consumer "durables," as economists call them—to "offer a reasonable expectation," in the judgment of bankers, "of lasting the life of the loan."[119]

Besides credit transactions, the second and most important factor behind the new complexity and tension surrounding car buying was the trade-in. Until about 1920, car buyers who already owned a vehicle generally had to sell it privately. New car dealers then did not accept trade-ins, in part because cars deteriorated so rapidly. As late as 1916, the Ford Motor Company decreed that no "second-hand cars" should ever be "brought into or stored in our buildings to be resold by or through us," and reminded dealers: "our business and that of our salesmen is to sell NEW Ford cars."[120] By the 1920s, however, the emerging replacement market, along with the increasing durability of cars and availability of credit, forced Ford and other manufacturers and their respective dealers to accept used cars in partial trade for new ones. The pressures for doing so increased as new car sales went flat as early as 1924, and by 1927 when for the first time the number of used vehicles sold exceeded the number of sales of new cars.[121] By1930 the industry had what *Fortune* magazine called a "used-car problem," as old vehicles had started "to back up on the new-car market like a clogged sewer."[122]

For car buyers, the value of their old vehicle as a trade-in could become a hotly contested topic of dispute. Salesmen at car dealerships sometimes practiced "plate-glass appraising," putting a value on the buyer's car just by glancing at it through the showroom window, but even more methodical assessments of a prospect's car's value could generate conflict.[123] Because dealers never knew in advance exactly how much they could get when reselling a car, they tended to undervalue trade-ins. But as they usually had little room to lower the price of a new car, the only way to close the deal was often to overvalue the trade-in, knowing that the larger the "allowance" they gave on the trade-in, the less profit they made on the overall transaction. When Chrysler and Plymouth salesmen were asked what "most they would like to know about selling automobiles," a sales manual reported that the following replies recurred again and again:

> "How can we make a prospect realize the true value of his old car?"
> "How can we sell a prospect on a fair and just appraisal?" "How can we prevent a deal from going on the rocks because a prospect wants more for his car than it is worth, or because he has received a higher bid from a competitor?" "How can we tactfully show a prospect that his car is now worth only a fraction of its original price?"[124]

For their part, buyers regularly overestimated the worth of a vehicle and often misrepresented its condition as well. In short, "used cars are what forced automobile retailers to be car *dealers*," concludes historian Stephen Gelber in his study of automobile selling.[125] Lest they be taken advantage of, buyers also had to be ready to "deal," but dickering over the value of trade-ins injected more contentiousness and stress into the regular if infrequent ritual of replacing old cars with new ones.

If haggling was the price of automobility, consumers were more than willing to put up with it, as cars remained the most alluring and desirable of possessions. By 1925 some 17 million Americans already owned one, and the number would keep rising relative to population nearly every year up to World War Two, except during the Depression years from 1931 to 1936.[126] Few people yet questioned the environmental or social implications of personally owned motorized transportation, and as long as Americans could get credit, they eagerly kept driving and buying new automobiles. One of those enthusiastic owners was banker William Ashdown, who in 1925 published an article, "Confessions of an Automobilist," meditating on some of what he saw as the deleterious consequences of the automobile's mass adoption.[127]

Ashdown recounted his own experience with cars and analyzed the effects he saw motor vehicle ownership having on people's pocketbooks, psychology, and social behavior. "The lure of the car," he suggested, broke down people's "sense of values" and left them "quite helpless." The "psychological processes of car owners are much alike" and had evolved according to a predictable pattern. "First you want a car, and then you conclude to buy it. Once bought, you must keep it running. . . . Therefore you spend and keep on spending, be the consequences what they may. You have only one alternative—to sell out; and this pride forbids." Ashdown claimed never to have "known a man to give up an automobile once owned, except to buy a better one."

Not wanting readers to think him a "tight wad," Ashdown explained that he was simply "thrifty" with money, "at least until I became a motorist." Before getting a car his "greatest single extravagance" had been the purchase of a bicycle. Around 1917, however, smitten by the lure of the car, he bought "a modest secondhand machine," paying in cash the equivalent of a tenth of his annual income for the automobile, far more than he had ever before spent on anything in his life. Unlike with his bicycle, which had "carried with it no upkeep or heavy depreciation" and whose "first cost was its last," with the automobile "the first cost was decidedly not the last." He recited the litany of expenses familiar to every owner: "Maintenance and upkeep," gas and oil, the five dollars a month to garage the car, unexpected troubles and minor necessities, along with "several accessories that added to the appearance." There were also some totally unexpected expenses, such as those he incurred driving his non-car-owning friends around. Finally, as a "green," or inexperienced, driver Ashdown had a few accidents, all of which added to the costs of keeping an automobile. Lacking "a mechanical turn of mind," as he put it, he always paid others to repair his car. He "carefully set down all expenses" connected with the car, "curious to see if my budget was working out, but the figures mounted up so fast that I dared not look the facts in the face and so closed my books."

His first car "taught me these truths," he confessed. "A used car is a risky investment unless its history and previous ownership be fully known," and "a cheap new car is a better hazard than an expensive used one." This realization allowed him to rationalize the purchase of his second vehicle, which he bought new. "Although the spending of one tenth of my yearly income on the old car was a severe shock to my conscience," he admitted, "I found no difficulty in consenting to spend a quarter's salary on a new one. I had surrendered completely to the automobile." Ashdown was very familiar with such behavior, for numerous clients at his bank had sought loans to buy a new car when they already owed money on another vehicle, on their home, or on a business they had started. "The ease with which a car can be purchased on the time-payment plan is all too easy a road to ruin." Many people "were car poor" and "lived on the brink of danger," he asserted, saving nothing and becoming susceptible to being pushed over the edge by any unexpected emergency.

Overall, "the automobile stands unique as the most extravagant piece of machinery every devised for the pleasure of man," Ashdown concluded. "But—I still drive one myself. I must keep up with the procession, even though it has taken four cars to do so."[128]

What did keeping up mean for Ashdown and others who adopted these extravagant machines? It obviously meant buying and continuing to buy automobiles. Ashdown bought his first after friends kept telling him that, having just won a good job and "arrived," he should signal that success through car ownership. Thus, the lure of the automobile was partly social, a desire to measure up to other people's expectations. As a car owner, Ashdown became more fluent in the language of "Motor," learning the different makes of vehicles and the subtle nuances of social meaning their possession conveyed. He came to realize that his first automobile did not measure up to people's expectations as to what a person in his position should be driving, so his second "looked like a banker's car." As a new vehicle, it bore none of the signs of stylistic obsolescence that industry designers were exploiting to encourage consumer dissatisfaction and cultivate the desire for the new. Ashdown does not say what motivated him to trade in his second car for a third and then for a fourth, the one he was driving when he wrote his article. But he implied that to keep up with the "procession" he would want his car to remain fashionable and not cost too much to run even though all automobiles have "an insatiable appetite for money." He, like most who become accustomed to the convenience of car ownership, wanted to be able to jump into his vehicle at any time, have its motor start on demand, and be able to take off as personal whim or necessity dictated.

The actual need for or utility of an automobile was something Ashdown only begrudgingly acknowledged. "A car," he observed, "was considered distinctly a luxury" when he first bought one, "not, as at present, a rudimentary necessity," at least for the "carpenters, masons, bricklayers, and so on, living in inaccessible suburban places, who used their cars—all new ones—to go to and from their work." He lamented the passing of "the humble bicycle or the trolley car" and how "walking to-day is a lost art." Yet an increasing number of Americans, including many who did not dwell in "inaccessible suburban places," just could not keep up economically unless they owned or had access to an automobile. This imperative only grew stronger as trolley lines were abandoned. The technology that Ashdown considered an intoxicating extravagance remained extravagant but had become, in barely a generation, a household necessity for many.

Whether an extravagance or essential, the automobile was a complex machine to operate and run. Millions like Ashdown had done this, however, starting as total novices confronting the confusing panoply of controls that started, steered, and managed an automobile under way. It is to those challenges of operation, the necessary work automobilists had to perform behind the wheel, that we turn in the next chapter.

Chapter 3

Running a Car

A FULL-PAGE PHOTOGRAPH IN THE 1920 BOOK *Everyman's Guide to Motor Efficiency* depicted what a driver at the time would have seen on climbing behind the wheel of an automobile. No machine had ever presented consumers with a more daunting, less friendly interface, to use the now familiar term for the place where humans interact with machines, and the controls they employ to achieve that interaction. "This is where the salesman leaves you after you have bought the car," the caption reads: "You are confronted with dials, gages [sic], levers, pedals and switches used in the control of the car."[1] To men and women who had never run a device more complicated than a bicycle, a lawn mower, or sewing machine, getting into the driver's seat of an automobile could produce, as the caption suggests, a feeling of dizzying confusion. Even the simplest cars had, along with their steering device, controls to regulate the motor as well as for stopping and starting the vehicle; more complicated automobiles, like the "very large and powerful" Chadwick, had "no less than ten components to the control system," explained one automobile handbook. Because of such complexity, "there will always be some people who will not and cannot learn to drive," the author concluded.[2] This was especially true during the first two decades of the century, and many decided, as had the character named Aunt Hattie from the story "Edith Burnham" (discussed in chapter 1) regarding sewing machines in the nineteenth century, that they could not learn enough to meet the demands of the latest technology.

Driving a car has always required some practice and the acquisition of new skills, and beginners invariably feel awkward, uncertain, and even scared. Because of the speed at which automobiles move, the risks of death or grievous injury are compounded relative to most other consumer technologies, and an error or lapse on the operator's part can quickly lead to catastrophe. But early adopters faced even greater dangers running their machines in an operating environment constructed for the horse and buggy: single-lane dirt and gravel roads, tight corners where one could not see oncoming traffic, and rickety bridges barely as wide as a car. The propensity of early automobiles to skid, a result of the smooth, tread-less tires of the era running on dirt and gravel roads, only added to the dangers. But first-time motorists found their greatest challenge learning to manipulate the many controls of their machines. This was especially the case with gasoline automobiles, the dominant type from around 1900 well into the 1920s that would demand so much of users compared to later vehicles.

Early adopters registered this challenge in their rhetoric, commonly referring to horseless carriages as "machines" or "motors" and borrowing other terms from the language of machinery and factory, as if horse-era language were no longer useful. Adopters "ran" or "operated" automobiles, verbs bespeaking the intimate involvement they had with the mechanism of their vehicles. They received "operator permits" and "operator licenses" from the states and municipalities that regulated the use of automobiles.[3] Organizations like the "Massachusetts Automobile Operators' Association," an early promoter of highway safety, similarly adopted a usage derived from industrial rather than equestrian experience.[4] So, too, did new coinages such as *autoist* and *automobilist* suggest the belief of early adopters that the word *driver* did not adequately describe their work.[5] Many also rejected the descriptor *motorist* because of its similarity to *motorman,* used for the working-class operator of a trolley car. It would not be until the 1930s, when automobiles had become both ubiquitous and their operation easier, that the two words—the new coinage *motorist* and the older one *driver*—would universally be applied to operators of motor vehicles.[6]

Ironically, although early automobilists came to reject horse-era terminology, they considered horse-and-buggy experience valuable training for motoring and drew on it metaphorically in thinking about their new machines. A Lima, Ohio, operator derived a valuable lesson by comparing the strength and endurance of mechanized versus animal-powered vehicles. "There is no reason for compassion" when dealing with an automobile, he argued, for it being "a 'thing of steel' knows no pain and can endure treatment without injury which would ruin the animal made of *flesh and blood, like ourselves* [italics in original]."[7] Most motorists, however, advocated compassion for their cars no less than for beasts. "Don't drive your 'Oldsmobile' 100 miles the first day," exhorted a 1904 owner's manual. "You wouldn't drive a green horse 10 miles till you were acquainted with him. Do you know more about a gasoline motor than you do about a horse?"[8] Another writer pointed out that "a motor car requires almost as much sympathy as a horse if the best results are to be attained,"[9] while still another urged treating an automobile "as you would a good horse," giving "it loads that it can easily handle."[10] Yet another writer presumed human operators might instruct their machines no differently than they once taught their horses. Only the motorist who "has carefully trained his car and refrained from those excesses of speed and inattention which are deleterious," he believed, would get the most "service" from his vehicle.[11]

In short, "Auto Sense Is Horse Sense in a Mechanical Form," as another expert summed up the relevance of equestrian experience to automobile use. Operators of horseless carriages needed to be just as alert to and sympathetic to subtle changes in the behaviors and symptoms of their mechanical steeds as they once were to those of their four-legged ones. Such "sympathy" could enable them to avoid trouble before it happened.[12] The skilled operator "feels the road through his wheels as the horseman feels his horse's spirit through the reins."[13] Just as the driver of a horse and buggy monitored his animal's fatigue and thirst, the motorist must pay attention to the "thirst" of his automobile—its need for water, fuel, and oil—along with its other vital signs such as engine temperature and tire wear. Like horses, automobiles had needs, and compared to previous consumer technologies like sewing machines or bicycles, the new technology was needy in the extreme.

The big question facing novice autoists was how to translate old-fashioned horse sense into the knowledge they really needed of the new, mechanical form the horseless carriage presented. They certainly did not require a "first class mechanical training" of the sort one got as an apprentice in a machine shop, asserted a 1912 technical handbook.[14] "What you want," explained the writer, who coined the aphorism that auto sense is simply horse sense modernized, "is to know your car like you knew your old family horse," that is, limb by limb and from nose to tail.[15] Operators should therefore familiarize themselves with everything from the vehicle's radiator to its gas tank, from its spark advance control to its carburetor, and from the brakes to its gears; they should not only know what these were and where on the vehicle they could be found but also have some sense of how they functioned. Because operators would invariably have car trouble, such knowledge was deemed necessary to diagnose the causes and to effect an adjustment or repair. Unless owners paid a chauffeur to take responsibility for such problems, they constituted their own first line of tech support when car trouble occurred, a subject we take up in the next chapter. But first we here consider them as operators, examining how car buyers learned to run their complex machines and the knowledge and skills they required for such work.

Early automobile buyers learned the rudiments of operation in three ways or some combination of them. Some attended the "schools" that cropped up to teach beginners. Others taught themselves through observing others and by trial and error. Most, however, learned from the dealers who sold them their cars, just as nineteenth-century purchasers of sewing machines had once been taught either by company demonstrators or door-to-door salesmen. "Driving instructions were as much a part of every Model T sale as the engine," recalled a Ford dealer, looking back on his early career in the 1910s. "Often, a new owner demanded that I teach his wife and every kid old enough to reach the pedals how to drive."[16] Such instruction was usually "in your own car" and after you had paid for it, noted a handbook author, because dealers knew that novice operators frequently made mistakes and crashed during their first lessons.[17] Such lessons were frequently brief and often ineffective. On taking delivery of a new Franklin in 1905, J. Walter Predmore, of Akron, Ohio, received "about an hour's lesson" from the dealer.[18] Myron Stearns, who bought his first car used in 1910, received an even shorter lesson. After he signed the deal, a mechanic drove the car "out onto a country road for me and put me behind the wheel," Stearns reported.

> He told me the car was a roadster. I knew that. He said it had a planetary transmission. I tried to look intelligent. Then: "Push that pedal with your foot," he said. We went forward. "Now push this pedal," he said. We stopped. "Now push the third one." We went backward. "That's all there is to it," he said, "except high speed. Here's the switch

to turn off the motor when you're through." I had learned to drive! Exactly one minute![19]

Although Stearns surely exaggerated, driving instruction would long remain similarly haphazard and unregulated.

Nevertheless, a small if unknown percentage of early motorists learned to drive in classes that anticipated modern driver's education. As early as 1903, "the most important attempt to provide formal training for motorists and mechanics" began at the Boston YMCA, according to automotive historian James Flink.[20] YMCA instruction spread to other cities, and by mid-1905 over two thousand students had enrolled in the organization's automobile courses nationwide. By 1908, nearly that number had completed YMCA courses in New York City alone. Enrollment statistics there from 1915 show that car owners and prospective owners made up between 25 and 45 percent of the students, while the rest were comprised of professional garagemen, dealers, chauffeurs, and mechanics.[21] YMCA courses soon had competition, as any individual with a car and entrepreneurial spirit could set himself up as a driving school and start teaching students. Little information survives on such institutions, and only after World War Two, spurred by legislation and insurance companies, did driver's education gain a firm toehold in the public schools. As late as 1931, half of all states did not even require a license to operate a motor vehicle, let alone any previous instruction.[22]

Until after the Second World War, then, most people learned to drive from friends and through self-instruction. In the 1910s, Marjorie Sweet, "leaning from the seat next to the driver's seat" of a friend's one-cylinder Cadillac, first practiced the different tasks required to drive: steering, actuating the brakes, and changing gears.[23] Eventually, Sweet soloed, bought her own automobile, and continued to learn. Elsie Bossert had an even less structured experience in rural Indiana during the 1920s. "We didn't have any training," she recalled of learning how to drive, "only what you would pick up from your parents or somebody. You would go out in a field, as a usual thing, and drive around the wheat shocks to learn to drive and to miss them."[24] Scholars call this approach learning by using, and while in the case of automobiles it invariably produced close calls and even accidents, for better or worse it is still characteristic of the way we learn to use most personal technologies.

Learning by reading is another common approach to new technologies, and early adopters often turned to the owner's manual that came with their vehicles, as had purchasers of sewing machines and other devices before. Additionally, they consulted a variety of other how-to-do-it books and articles about motoring. From about 1900 to 1930, newspapers and magazines, including family periodicals like the *Saturday Evening Post* or *Literary Digest* and women's magazines such as the *Ladies' Home Journal* and *Good Housekeeping*, also published pieces about driving and caring for cars. All such publications are helpful to us today in reconstructing the tasks and procedures operators had to master in order to operate early automobiles.[25]

MUCH OF WHAT USERS MUST DO to successfully operate an automobile (or any other technology) is determined by a machine's design and construction, by the way its interface is set up and how it works. Users may have considerable choice as to what they do with their machines, in the case of automobiles whether they drive to the country, cruise around town, or just park and watch the sunset, but they have little choice as to *how* to make machines follow their wishes. The procedures for starting a car's engine, putting the vehicle in motion, slowing it down and stopping, or replenishing its oil, water, and gasoline—these tasks must be performed in prescribed ways. The automobile largely determines how and when such things must be done. Everybody who has the same make, model, and year automobile, regardless of their age, sex, race, or class, essentially must meet the same technical demands from their machine. Historians have long been skeptical, and rightly so, of "technological determinism," the idea that behaviors or events are determined by machines. The skepticism is justified, given the uses to which people put automobiles and the fact that technology alone clearly does not itself cause broad social and cultural developments and behaviors. But once designed and manufactured, machines *do determine* what behaviors users must employ as operators, that is, in their communication with the machine at the technology's interface.[26]

In operating their cars, early consumers had to master five tasks that are universal to running an automobile and that can be called the five *S*'s: starting the car; steering the vehicle; selecting the appropriate gear;

stopping or slowing it; and finally, supervising the machine while under way and meeting its changing needs.[27] By considering each of these tasks we can understand what was involved in running an early gasoline automobile and how, over the period from 1900 to the 1930s, these tasks changed. Some were drastically simplified while others were eliminated by automation and other technical changes, which typically eased the work of operating but made the machines themselves more complicated.

Automobiles, like any machine not powered by wind, water, or muscle, have to be started or turned on. As explained in chapter 2, automobilists who owned electrics did this by literally flipping a switch, while those with steamers and internal combustion engines confronted more complicated starting procedures. With a Stanley Steamer, drivers first had to make sure the pilot light, fueled by gasoline, was burning before lighting the main kerosene burner, which heated water in the boiler to produce steam. After about thirty minutes, their steam pressure gauge would tell them when they had a sufficient head of steam to get under way.[28]

Owners of virtually all gasoline cars before 1912 participated in a less time-consuming but similarly complicated, and often more arduous and hazardous starting ritual. Before hand cranking their engines, they often had to work a hand pump on the dash, designed to build up pressure in the fuel tank in order to force gasoline to the carburetor. Additionally, operators manipulated a manual choke that adjusted the fuel-air mixture of the carburetor for starting, a step still required with the small gas engines found today in many lawn mowers, leaf blowers, and chain saws. Owners of automobiles with one- or two-cylinder engines sometimes had to prime them, or supply each cylinder with fuel, before firing the engine.[29] And virtually every operator of a gasoline car also had to manipulate two other controls, usually located on the steering column, before getting to the work of cranking. First, they set the manual throttle, or "speeder," that controlled engine speed, for early automobiles customarily did not have accelerator pedals; then they moved a lever that retarded the "spark," that is, delayed the ignition of the gasoline-air mixture in the cylinder until late in the power stroke, helping prevent potentially dangerous backfiring that might turn the crank violently backward during hand cranking (once under way, motorists advanced the spark so as to fire cylinder earlier in the power stroke, providing more power and saving fuel). With everything ready for starting, motorists then approached the dreaded task of manipulating the starting crank. Usually this meant getting out of the driver's seat and going to the front of the car, where the crank protruded through the radiator, although on some early runabouts with engines beneath the operator's seat, the crank extended through the side of the vehicle, allowing the operator to reach it from a seated position.[30]

Cranking the one- and two-cylinder cars of the early 1900s was relatively easier than cranking those with four or more cylinders, but cold weather caused engine oil to become more viscous and made any engine harder to start.[31] Furthermore, all internal combustion engines were susceptible of backfiring when being started. Before the advent of the self-starter in 1912, over a third of all automobile injuries were "cranking accidents" where fingers, wrists, and arms got in the way of rapidly spinning cranks.[32] One motoring periodical referred to "Chauffeurs' Fracture," a "peculiar and complicated fracture of the bones of the forearm" that showed up among those who carelessly cranked gasoline engines.[33] Because of these dangers, right-handed motorists were advised to crank with their left hand so that in the case of injury they "can get along by using the other hand."[34] They were also urged to place all of their fingers, including the thumb, on one side of the crank's handle so, if the engine did backfire, flinging the handle in the other direction, the handle would leave their hand without trying to take the thumb with it. At least one person allegedly died from cranking: Byron Carter, an experienced auto industry executive, was hit in the head by an errant crank while helping a motorist start her stalled car. The tragedy is said to have inspired his friend, electrical engineer Charles Kettering, to develop the electric starter, although everybody sufficiently hated hand cranking that no additional incentive was required to find another way to start a car's engine.[35]

Under certain conditions, operators might avoid cranking by starting "on the spark," as the technique was called. It was most likely to work with cars having four or more cylinders that had recently been running, but it also required a bit of luck. When an engine was shut down, sometimes it stopped with one of its pistons at the peak of its compression stroke, having compressed a charge of gasoline and air. When this happened, the next time the ignition was turned on, a spark in that cylinder just might ignite the charge, pushing the piston down, turning the crankshaft, and thereby starting the engine. The longer an engine had not been running, however, the less likely it would start on the spark because the compressed air and gasoline would have leaked out around the piston rings, spark plug, or cylinder head.[36]

Since the advent of internal combustion engines, many had dreamed of automating the arduous and risky cranking required at start-up. As early as 1911, Winton buyers enjoyed self-starters that utilized compressed air to crank their engines. The car's engine when running pumped air under pressure into a reservoir, where it was then available the next time for starting the car; an additional benefit of the system was that owners could inflate flat tires without arduous hand pumping.[37] Inventors also experimented with other kinds of devices to take over the work of cranking, including spring-driven mechanisms as well as motors driven by acetylene gas; a few inventive individuals rigged ropes and pulleys so they could at least crank their engines while sitting in the driver's seat.[38]

Starting in 1912, however, some motorists glimpsed the future in the form of the electric, or self-, starter. Developed by the aforementioned Charles Kettering for General Motors, the system debuted that year on the Cadillac, GM's top-of-the-line vehicle. Electric starters are the modern means of starting a car with an internal combustion engine, allowing operators simply to depress a button, causing a small but powerful battery-driven electric motor to turn over the engine. There were numerous consequences of self-starting, one of which was to deprive the electric automobile of its primary selling point. Another was that once automobiles were equipped with electric starting, which included a storage battery as well as a generator to recharge the battery, all manner of automotive electrical accessories became possible, including electric lights, windshield wipers, and eventually radio. Although some men in the auto industry claimed that large numbers of women drivers was another consequence of electric starting, historian Virginia Scharff has shown that by no means were women deterred from operating gasoline cars by the fact they had to crank their engines. Moreover, it is simply "auto industry folklore," she concludes, that men enjoyed cranking and developed self-starting for women's benefit.[39] Nobody liked the chore of hand cranking, and as the young Long Island artist Hilda Ward explained in 1908, if "doing a little extra cranking" was the price one had to pay for the mobility and convenience of automobiling, so be it.[40]

In spite of the invention of the electric starter, millions of automobilists continued to crank their engines for another ten or fifteen years. It is misleading to claim, as has one historian, that the electric self-starter had come "into general use" by 1913, when "nearly fifty other manufacturers adopted the innovation."[41] That year, and well into the early 1920s, the majority of registered cars on American roads still lacked electric starters; in fact, Ford sold as many or more Model Ts than all other cars produced by American automakers combined, and until 1919, when the feature became standard equipment on Ford coupes and sedans, they all lacked self-starting. Only in 1926 did electric starters become standard on *all* Fords, so not until some moment in the mid-1920s did this user-friendly feature become the dominant means of starting a car.[42] Even in the 1930s, after electric starting had become universal on new cars, Fords and other vehicles still made provision for hand cranking and supplied cranks in the car's tool kits for emergencies.[43]

By the early thirties, however, hand cranking had become unusual enough to attract attention and provide grist for stories. One tale, related to me by a man who was a child then, involved his neighbors, an elderly couple that lived down the street in Scarsdale, New York. Every morning, the pair would emerge from their house and walk to their car parked on the street. The husband would climb behind the wheel while the wife would proceed to the front of the car and furiously commence cranking. The pair would exchange animated and often angry words before the old car's engine clattered to life. At that point, the wife would jump into the passenger seat next to her husband and the couple would drive off. Neighbors found this daily ritual amusing but puzzling. Why, one of them finally asked the husband, did he make his wife do all the cranking? He answered by saying that the hardest part of starting an engine was not the cranking but rather the delicate manipulation of the spark advance and throttle levers. Only he had the mechanical knowledge and sensitive touch to handle that task, he claimed; cranking anybody could do.[44] The story not only inverts the usual gender stereotype about hand cranking but also reminds us how slowly older technologies disappear from day-to-day experience. The Depression, which caused many to sell cars they could not afford to keep on the road, followed by wartime gasoline rationing and scrap metal drives that caused the junking of many more hand-cranked relics, finally eliminated a long familiar but frustrating aspect of automobile operation.[45]

STEERING WAS THE SECOND OF THE FIVE S's operators had to master in order to run an automobile. It was an experience foreign to many in the nineteenth century, when most forms of transportation did not need to be steered. Trains and trolleys simply followed the tracks, while buggies and animal-drawn vehicles relegated steering to a collaboration of driver and animal. Horses were intelligent enough to follow a road or trail and required only an occasional tug on the reins to make a sharper turn left or right. For millions of people, their first experience with steering would have been on a safety bicycle in the 1890s, guiding it by turning the handlebars. Also in the waning years of the century, members of the middle classes increasingly owned small pleasure boats on lakes or the seashore. Most of these were steered by tillers, although a few had wheels like large ships.

A lack of familiarity with steering with a wheel, then, may help explain why so many early automobiles employed tillers for the purpose. The steering wheel "has met with such general disfavor in this country that levers are used almost exclusively," noted an electric vehicle handbook, justifying tiller technology.[46] The ten thousand men and women who bought a curved dash Oldsmobile between 1900 and 1905 and made this the most popular American car before the Ford Model T steered it with a tiller. Whether one sat behind the tiller, as in an Oldsmobile or electric vehicle, or in front of the tiller, as in a boat, to turn to the right one moved the tiller to one's left, and vice versa; the device and its motions were quite familiar.

Whatever the means, steering was "one of the first things to be learned by the novice," explained an automotive veteran in 1908. "Learn to steer before attempting to learn even the most fundamental things about controlling the power plant," he suggested.[47] The first generation of automobilists found steering so challenging that they often practiced it separately from operating the vehicle itself, sitting in a nonmoving car whose wheels were jacked up off the ground. Hilda Ward described attended a driving school in New York City in 1906 that used such pedagogy. In a car raised off the floor, she manipulated the steering wheel and worked the brake under the critical eye of the instructor who sat next to her barking out warnings: "Chicken on the right, baby-carriage on the left, trolley-car ahead!" Such training "at the end of a week," she quipped, left the student "incapable of running a wheel-barrow for the visions I should be having [of hitting things]." Even though the exercise made her anxious, she claimed it let her "get familiar with the steering-gear and brakes by the easiest and most natural process."[48] The technique enabled students to learn "how the manipulation of the steering wheel affects the direction of the car," another supporter explained, for in a stationary vehicle raised off the ground trainees could readily observe how turning the steering wheel moved the car's front wheels.[49]

But the method would not teach two other crucial lessons about steering early automobiles, both of which required hands-on experience in machines actually moving on roads. The first lesson was that any movement of the car's front wheels caused by hitting a rock, say, or encountering a pothole, was promptly transferred to an equivalent movement of the steering wheel. This was a design shortcoming of early steering mechanisms and was one reason why how-to-drive books stressed a correct seat, or posture, and steering wheel grip for operators—back straight, torso erect, and with two hands holding the wheel (though not a "fierce grip"). The propensity of the steering wheel to be jerked out of the operator's hands if the front wheels hit an unevenness in the road was one reason for experts condemning as "bad method" any "careless lounging" or "nervous, uncomfortable" position in the driver's seat.[50] With the introduction in the 1910s of so-called "irreversible steering," where movement of the steering wheel turned the front wheels but not the reverse, operators no longer had to worry about the steering wheel (by then universal) being torn from their grasp.[51]

The second lesson motorists would pick up only through actual operating experience was how much physical effort it took to turn the steering wheel when a car's wheels were on the ground. Automobiles in our period did not have power steering, but for a number of technical reasons that lasted through the 1920s, turning the front wheels took much more effort than it would in later vehicles. One reason was that engineers designed early cars with relatively low steering ratios, a measure of how many turns of the steering wheel is required to turn the front wheels from full left to full right position, or lock to lock. Today, a car without power steering might have a steering ratio between three and four, but well into the 1920s automobiles commonly had ratios of one and a half to two. These provided "quick" steering, that is, operators while under way could avoid a rock or pothole just by turning the steering wheel very little. The corollary of quick steering, however, was that the low ratio reduced the mechanical advantage and it took relatively more strength to turn the steering wheel. Given that the wheels of early cars tended to be of large diameter and that vehicles were heavy, turning the steering wheel to move those wheels, especially when parked or moving slowly, could be like "turning a tree in the ground."[52]

The strength, coordination, and strangeness steering presented to early adopters prompted a debate in the pages of the *Journal of the American Medical Society* in 1910, as physician-motorists pondered the question of whether steering would ever become automatic, something operators could do without thinking, or would always demand unfettered mental and physical concentration. While a minority were skeptical and believed it always would remain a challenge, most thought that as people became familiar with steering wheels and their operation, the work would indeed become automatic.[53] This position would eventually be confirmed, and the idea that steering was a major challenge was forgotten.

The third *S* was shifting. The "Gear-Shifting Bugaboo," as it was called, was all but unavoidable unless one bought an electric or steamer.[54] Cars powered by internal combustion needed gears because their engines, unlike steam engines or electric motors, only developed acceptable levels of power at high revolutions per minute. Gearing enabled such engines to run fast while the car moved relatively slowly. Before the introduction of the first automatic transmissions in the late 1930s, shifting gears generally meant using a foot-activated clutch while manipulating by hand a "change gear," or shift lever. Not only could this be physically demanding, but it also required precise coordination of hands and feet, especially when starting from a standstill so as not to stall the engine. Moving forward from a dead stop on a an upward incline posed the greatest challenge to the motorist's coordination, for stalling the engine under such circumstances could endanger the operator, his or her passengers, and others should the car be allowed to roll backward.[55] "Handling of the ordinary pedal controlled clutch is an occult science," one motorist said of the technology in 1905, "known only to the initiated and only to the learned in the school of experience."[56] Whether on a hill or flat terrain, to the early 1930s and beyond, using the clutch and changing gears remained a source of continuing consternation for many.

Another early worry was "stripping" the teeth from transmission gears as a result of poor shifting technique.[57] Even though the auto industry improved its metallurgy and by the 1910s was making gears out of robust steel alloys, until the development of synchromesh transmissions in the 1930s, which synchronized the speeds of mating transmission parts before they engaged, motorists commonly ground or clashed gears when shifting, especially when downshifting from a higher to a lower gear. To downshift before synchromesh, operators could not simply disengage the clutch and move the shift lever out of one gear and into another; they had to employ a technique called double-clutching.[58] This entailed first disengaging the clutch, usually a foot pedal, and almost simultaneously moving the shift lever out of third or high, say, into neutral. At that point motorists released the clutch and then quickly pressed the pedal a second time, at which point the shift lever was moved from neutral into the desired gear, here second or intermediate. Releasing the clutch the second time completed the change-gear operation, save for one step that has been omitted. As operators moved the shift lever from neutral into the lower gear, they also advanced or "blipped" the throttle slightly so as to bring the engine's speed up to the somewhat faster speed of the transmission. Some skilled motorists could coordinate all of these steps without clashing gears, but many motorists found smooth and noiseless shifts hard to achieve.

Many detested the combination of clutch and sliding gears, not least Henry Ford, who referred to the arrangement as "crunch gears" and is said to have never learned to drive a car so equipped.[59] In his Model T, he utilized a different arrangement, a so-called "planetary gear" transmission, which eliminated the need for double-clutching and the tricky synchronization it required of the operator, with one hand on the shift lever, another on the steering wheel, one foot on the clutch, and the other foot the accelerator. Instead, Model T operators shifted with their feet. Two of three pedals on the floor of the driver's compartment controlled the transmission. Pushing the leftmost of the three pedals to the floor put the car in low gear; letting it up halfway moved the transmission into neutral; releasing it altogether placed the car in high gear, or direct drive (there was no intermediate, or "second," gear). The rightmost pedal braked the car but by acting on the transmission rather than the wheels. To shift into reverse, one pressed down on the middle peddle.[60] Because Model T transmissions were always engaged unless operators held them in neutral with their feet, some humorous and even dangerous situations resulted.[61] "Often," essayist E. B. White observed, fondly recalling his own experiences crank starting a Model T, "if the emergency brake hadn't been pulled all the way back, the car advanced on you the instant the first explosion occurred and you would hold it back by leaning your weight against it."[62] It was as if the Ford had a mind of its own, White explained. "There was never a moment when the [transmission] bands were not faintly egging the machine on. In this respect it was like a horse, rolling the bit on its tongue."[63] Millions accepted this peculiarity of the Tin Lizzie, finding it a small price to pay for being freed from the gear-shifting bugaboo.

Another complexity with early sliding gear transmissions was the absence of standardization in the shift pattern. The now long familiar H shape through which operators shifted gears became nearly universal by about 1910, but the position of the gears within the shape varied considerably. Victor W. Page's 1917 volume, *How to Run an Automobile*, described at least three variations: in makes such as the Cole, whose transmission exemplified one variant, operators would find first gear at the top, leftmost position of the H; the Dodge and other makes required a second shifting path wherein operators located first gear at the top of the H but on the right side; while the Packard Twin Six typified a third variation in which operators obtained first gear by moving the shift lever to the right but down, to the bottom of the H. Shifting into second, or intermediate, required operators of the Cole to simply move the shift lever straight down; in the Dodge or Packard Twin Six, however, operators moved their shift levers first up or down to the neutral position in the center of the H and then sideways before finally moving it down (Dodge) or up (Packard) into second gear. Going from second to third, or high, gear similarly was different: in a Dodge or Packard Twin Six, operators moved the levers straight down or up one side of the H, while in a Cole they would first move the lever up, then transit the middle part of the H, and finally push the lever up into high. Reverse gear was similarly scattered around the

H, whether on the bottom left, as in the Dodge and Cole, or on the upper right, as in the Packard Twin Six.[64]

A beginner learning to run an automobile for the first time would hardly care what the shift pattern looked like: the one in his car would seem normal enough. Problems arose, of course, when operators accustomed to one pattern drove a vehicle having a different if not completely contrasting arrangement. In fact, even though manufacturers always labeled the position of gears, either with numbers and letters on the gearshift knob or on the transmission cover, these variations in shift patterns provided a recipe for confusion and disaster. Critics early bemoaned the "crying need for the standardization of control elements," but only in the late 1920s did the industry finally standardize the gear locations within the H pattern, thereby contributing to motoring safety.[65]

Whatever the pattern, motorists disliked shifting gears so much that many avoided it whenever possible. Some apparently never used reverse on their cars, the gear most difficult to engage, and carefully parked so they could always get out without backing up. Among these were Gertrude Stein, the writer, art patron, and expatriate, along with Juliana Force, one of the founders and benefactor of New York's Whitney Museum of American Art.[66] Wealthy owners even installed turntables in their residential garages so they could drive in and out in forward gear; once their car was in the garage, they spun the turntable so that the car would face out for the next trip. More common was the practice of ignoring the low and even intermediate gears on automobiles and driving all the time in high. This was possible in the first two decades of the century especially because early engines provided far more torque at low rpm than is common today, so motorists could readily start their cars from a standstill in second or even high without stalling, especially if they "slipped" the clutch—that is, disengaged it partway. Experts criticized relying "mainly upon the clutch to control the speed of the car," because it rapidly wore out the unit.[67] Yet car salesmen touted the ability of their cars to "creep along in high," and owners simply exploited that possibility, ignoring the crunch gears altogether and driving around as if their automobiles had automatic transmissions.[68]

Meanwhile, engineers and inventors had been trying to simplify or eliminate shifting altogether. As early as 1905 a Boston firm offered the

Sturtevant "leverless" car, its primary selling point being the absence of a gearshift lever, but the machine was otherwise a commercial and technical failure.[69] The Model T was also "leverless," its planetary transmission actuated by foot pedals, and a number of automakers besides Ford also utilized that type of transmission in the early years of the auto age. Engineers in the 1930s came up with another innovation in the effort to simplify the shifting operation, which was called freewheeling. According to a historian of automotive technology, freewheeling "reduced the skill required to synchronize gear speeds" and avoid the clashing of gears by allowing operators to decouple the rear wheels from the transmission. First offered as an option by Studebaker in 1930, within a year about half of all American automobile makers adopted freewheeling, but the innovation had a serious drawback.[70] When a car was freewheeling, without the retarding effect of engine compression on its speed, only its brakes could slow it down. Because early brakes were often unreliable, many consumers did not wish to exchange the braking effect of their engines for freewheeling, so the innovation proved to be only a passing fad.

With the innovation of synchromesh transmissions, of course, operators avoided any clashing of gears when shifting between second and third or vice versa. Among the motorists who benefited from synchromesh were the eighteen thousand who bought a new, 1929 Cadillac, one of the first cars to incorporate the improvement. Operators still had to use their clutch, but they no longer had to double-clutch or synchronize the speed of their engines and transmission gears while doing so.[71] Chevrolet and other General Motors cars had synchromesh the next year, and by the middle of the decade virtually all automobiles incorporated the technology. Motorists loved synchromesh, as it not only removed much of the anxiety surrounding shifting but also enabled operators to easily downshift from third to second while descending steep hills, thereby slowing the vehicle and saving their brakes. And by the end of the 1930s, Oldsmobile engineers had fully automated the process of shifting, offering the first hydromatic, or automatic, transmission as an option on new cars.[72] It would take three or four more decades, however, before the majority of automobiles had automatic transmissions, finally relegating the gearshifting bugaboo to memory.

STOPPING ONE'S VEHICLE WAS THE FOURTH S that the early automobilist had to master. "The first thing on the novice's program should be stopping," recommended the author of a 1912 handbook, somewhat obviously. He suggested beginners initially "learn at once and remember for all time" the technique of "stopping the car by shutting off the engine"; only later, after acquiring "more skill in handling the car and increased presence of mind," should novices learn to use the brakes.[73] This advice not only acknowledged the shortcomings of early brakes but also the ways contemporary roads and tires contributed to the difficulties of stopping an automobile.

Until about 1920, virtually all cars had brakes only on their rear wheels, augmented by a safety, or "emergency" (now also called "parking") brake, which usually clamped on the car's drive shaft or transmission to slow the car.[74] Operators activated the brakes either by pushing on a foot pedal or pulling on a hand lever, and their effort was translated mechanically, through rods or cables, to the brakes. For a number of reasons, the brakes of this era were of limited use in stopping a car; one 1905 Winton owner thought so little of his that in a letter to the factory he suggested removing them altogether. "It does no harm" to leave the brake on the car, Winton replied, as "you might want it awfully sometime and if you didn't have it the chances are that you would never need it again."[75] Besides their limited stopping ability, early brakes were likely to cause a dangerous problem: the rods that activated the brakes invariably bent or got out of alignment, producing unequal braking on the left or right wheel, which often triggered dangerous skids. Brake cables also stretched unequally, producing similar results. "Skidding was so usual that no one even mentioned it," recalled one pioneer autoist, perhaps with some exaggeration, "unless the car made at least one complete turn."[76] When brake cables snapped, motorists found themselves without any stopping power at all. In short, early automobilists could not rely on their brakes, and as late as 1920 a guidebook advised them to drive in a manner so as to "use the brakes as little as possible."[77]

Another reason operators might avoid unnecessary braking was that it could damage their vehicles. A 1916 article, "Saving the Car by Careful

Driving," suggested there were "two kinds of driving: that which saves the car and its mechanism, and that, which, by its very conservatism, reduces the possibilities of mishaps to a minimum." The author believed "the two should go hand in hand" and that "the driver who is careful of his mechanism should be equally careful of human life."[78] That the author worried first about "mishaps" to the car and secondly about being "careful of human life" suggests the extreme fragility of early automobiles, especially their tires. The smooth, treadless, natural-rubber tires used on cars in the 1900s were their Achilles' heel, being prone to punctures and tears; they could also easily be destroyed by rapid acceleration, running over stones or curbs, or hard cornering. If the brakes locked the wheels, the ensuing skid might easily cause a blowout.

Given the limited effectiveness of early brakes, it is not surprising that some people got behind the wheel without having given much thought to stopping. An Indiana woman recalled an example of such behavior. Around 1920, her uncle was going to Indianapolis by train and had been asked while there to pick up a new car for a neighbor. He had never driven a car before, so on arriving at the Indianapolis dealer he received some rudimentary instruction, but the salesman did not tell him about the brakes and he did not ask about stopping the car. "When he got home he couldn't stop," the niece remembered, "so he went around the block and around and around and around, until he ran out of gas."[79] Another example, also from around 1920, involved a car owner named Emil who had a local mechanic chauffeur him around most of the time. One day, when the mechanic had warmed up the car, Emil unexpectedly drove off by himself. "It wasn't long," the mechanic recalled, "until I saw him drive by the garage in second gear. A few minutes later he drove by again and again and then the telephone rang." The owner's wife had seen him drive by and "was asking me to go out and stop the car for Emil. I jumped on the running board and yelled instructions that finally brought him to a full stop."[80]

Skepticism about brakes began to disappear during the 1920s, partly because public thinking about safety and motor vehicles underwent a sea change. People had been dying in automobile accidents since the first reported fatality in 1899, and the annual highway death toll had steadily mounted over the years. "Motor More Deadly Than War" was how the title of a 1927 magazine article put it, and it was this discovery that raised popular awareness of the carnage on the highways.[81] The horrific reality noted by the article was that the number of people killed each year on American roads regularly surpassed "the total death lists in the American forces during the World War." The media coined catchy terms for the phenomenon, inveighing against the "Deadly Driver" and "Morons on the Macadam" while clamoring that something be done about "Murder by Motor," "Motor Massacre," and the "Motor Menace."[82] The media's crusade did not itself make motoring safer, but it did help raise public awareness and bring into being a more safety-oriented climate.

Motorists in many states by this time were also encountering new and stricter licensing requirements, and some jurisdictions passed laws requiring drivers who lost their licenses for violating traffic laws to attend traffic school and then pass a special driving test to reinstate their license.[83] Also during the 1920s, states and municipalities, aided by federal funds, rebuilt the automobile's operating environment to make it safer. Engineers redesigned tortuously tight curves, replaced bridges that lacked guardrails, and eliminated blind curves, unnecessary intersections, and other highway features that, while tolerable in the days of animal-paced travel, had become places "where road wrecks happen" in the auto age.[84] Most visible to motorists were the many more miles of paved roads completed during the decade, almost doubling from 387,000 in 1920 to 694,000 by 1930; surfaced highways facilitated higher speeds while contributing to better traction and braking, less skidding, and even greater safety, all things being equal.[85]

Perhaps most significantly, during the 1920s brakes improved dramatically, starting with their being placed on all four wheels of a car. Previously most manufacturers thought it too complex and expensive to design front-wheel mechanical brakes because the front wheels had to turn to facilitate steering, but in 1924 Buick became the first major automaker to offer four-wheel brakes as standard equipment. When the low-priced Chevrolet announced that its 1928 models also would have come with four-wheel mechanical brakes, Henry Ford followed and made them standard on his Model A, introduced the same year. By 1930, all new cars had the safer four-wheel braking systems.[86] To be effective, of course, brakes had to be well maintained, and the propensity of mechanical brakes to need frequent adjustment prompted Massachusetts in 1934 to require owners to get their brakes regularly inspected and tested. The test required older vehicles with two-wheel brakes to stop, from twenty miles per hour, in forty-five feet, while those with four-wheel brakes had to stop in thirty. Neither standard was especially demanding, as some new cars required only twelve feet to stop from twenty miles an hour.[87] In fact, motorists with well-adjusted mechanical brakes could stop much more rapidly on paved road surfaces than they could on the gravel or dirt roads of the past, which helped to avoid treacherous skids.

"Modern" brakes, that is hydraulically actuated units on all four wheels, had been available on some luxury cars, such as Duesenbergs and Chryslers, since as early as 1921, but by the mid-1930s they were appearing on other makes. Chevrolet introduced the new technology on its 1936 models; as usual Henry Ford stubbornly clung to the old, switching from mechanical to hydraulic brakes only with his 1939 models.[88] With the new technology, no longer was the force with which motorists stepped on the brake pedal the lone determinant of the effectiveness of the brakes, as the operator's effort was now mediated and augmented by hydraulic fluid that reduced the physical effort required to stop a car by pressing on the pedal. More important, hydraulic brakes all but eliminated the problem, endemic with mechanical brakes, of unequal braking on one side or the other and the hazardous skidding that resulted.

Motorists also gained stopping power as automakers shifted to so-called balloon tires in the latter 1930s. Smaller in diameter, wider, and inflated to only about half the pressure of earlier tires, balloon tires were adopted for a number of technical and design reasons, but they also had an effect on safety. They put more tread area on the road than had earlier tires, and when used in combination with four-wheel hydraulic brakes and on paved road surfaces, enabled motorists to slow or stop their automobiles much more quickly and safely than ever before. A 1935 *Popular Mechanics* article asked readers, "Is Your Driving 'Touch' Modern?" The magazine's writers argued that recent automotive innovations such as balloon tires, hydraulic or vacuum-assisted four-wheel brakes (an early form of power brakes), synchromesh transmissions, and higher steering ratios (more turns, lock to lock) required motorists to "learn to drive all over again." The modern touch, the article claimed, involved a lighter and more delicate application of hand and foot to the controls. Take steering for example. After turning a corner in a modern vehicle, the magazine explained, a driver no longer should turn the steering wheel back to its original, wheels-straight-ahead position, as was done in an older vehicle. The higher steering ratio on new cars meant that, in making a right-hand turn, say, one turned the steering wheel at least a full revolution clockwise (the heightened ratio was required to give drivers enough leverage to easily turn balloon tires, with their larger footprint, especially when parking or standing still). Rather than return the steering wheel to its straight-ahead position, drivers should let it naturally spin back to its natural position on its own, "braking" the wheel it as it spins with pressure from a finger or thumb. The "importance of 'touch' in bringing out the best in the new cars," however, was "best illustrated" when slowing or stopping an automobile. "Any driver who is accustomed to 'jumping' on the brakes in order to bring his old car to a stop must turn over a new leaf the moment he takes the wheel of a new model," the article concluded.[89] With modern, four-wheel, hydraulically assisted, brakes it was easy to lock the wheels, which would cause the tires to skid and the car to lose control and stopping power.

Some motorists struggled to learn the "modern" touch required by new vehicles, but those driving older cars still needed a "pre-modern" touch. Technology demands that the skills and knowledge of operators is of the same vintage as their machines. In 1935, tens of thousands of Americans still owned and drove automobiles built in the early 1920s or even the 1910s. If lucky enough to own any kind of automobile, dust bowl farmers and black sharecroppers were groups whose vehicles were relatively ancient, requiring not a modern but rather an older touch. Even as late as 1937, half the cars on the road were built before 1933, and those included some 150,000 Model T Fords, the last of which was built in 1927, ten years earlier.[90] Operators of older machines continued to "jump" on their two-wheel mechanical brakes, double-clutch to shift gears, and hand crank engines to start them. These operators' habits, practices, and ways of thinking about their cars would become not just irrelevant but also dangerous when and if they moved to the latest model automobiles, as *Popular Mechanics* suggested, but until then a modern touch would not work.

ALONG WITH STARTING, STEERING, SHIFTING, AND STOPPING an automobile, operators have always had to perform another task, the fifth *S:* supervising their machine's performance while under way. Just as equestrians have always had to be sensitive to their animal's need for water or food, or to problems with its legs or feet, automobilists have had to monitor their automobile's supply of oil, water, and gasoline as well as the condition of its "shoes," or tires, and other components. The many maladies to which early automobiles were prone, however, made the operator's supervisory role more akin to that of a machine tender in a mill or factory than the driver of a horse and buggy.

The most insistent of an automobile's needs, of course, is its appetite for gasoline. Although cars in the early 1900s tended to consume no more gasoline per mile than do ours today, until the advent of filling stations around 1910 gas could often be hard to find and was frequently of poor quality. Some owners bought gasoline by the barrel and filled their cars in their home garages, carrying extra fuel in a can lashed to their vehicles while touring. If automobilists ran low on fuel while on the road, they might try to buy gasoline at a hardware store, feed store, mill, or other business that itself used gas and might be willing to sell a gallon. Transferring small quantities of the combustible liquid using pails or other containers almost guaranteed the introduction of dirt and other impurities, so experts advised motorists always to filter any gasoline they put in their tank. But pouring the combustible liquid through a funnel into tanks, with or without a cloth filter, risked causing a fire or explosion from built-up static electricity, as many users sadly discovered.[91]

Motorists in the early decades of the auto age also stayed busy meeting their vehicle's prodigious appetites for oil, water, and air. Along with gasoline, those vital fluids frequently leaked from engines, radiators, and tires due to the relatively primitive materials and seals used in automobile components, so operators ritually checked their supply, either every time before using their cars or on filling up their gas tanks. Leaks were exacerbated by the constant jolting cars got as they traveled over the era's poor roads, which could jar loose components such as the fuel hoses that connected carburetors with gas tanks. Supervising a car's gasoline supply took more effort than we might think because few cars had a gas gauge to show the amount still in the tank. Model T owners had it the worst: to tell how much gas remained in their tanks they had to get out of their cars, remove the driver's seat to gain access to the tank, and then unscrew the cap and peer in. Many carried a wooden rod for use as a dipstick to make a crude measure of the amount of fuel remaining, others carried a flashlight to illuminate the contents of their tanks, and a few tempted fate by striking a match in order to check the supply. On those early automobiles that did have gasoline gauges, technical reasons dictated that these instruments be located near the tank, which usually meant they were mounted on the exterior of the body, usually on the rear deck. Although the Chevrolet had such a gas gauge as standard equipment as early as 1914, Ford did not have one until it introduced its new Model A in 1928. Its gas tank was just forward of the dashboard, beneath the cowl, or rear portion, of the hood, allowing the gas gauge to be mounted in the modern position on the dash.

Operators checked their cooling water in the early 1900s the same way they would today, by opening the radiator cap and examining the supply—although recent automobiles seldom require fresh infusions. Back then, however, radiators routinely boiled over in hot weather or while climbing long hills. Model Ts were notoriously prone to overheating, lacking as they did any pump to move hot water from the engine to the radiator. Early water pumps themselves were balky and unreliable, further overloading automobile cooling systems. In 1909, Thomas Russell, the author of *Automobile Driving Self-taught,* implored operators to frequently check that their pumps were working. Your "next duty" after starting the engine was to observe the functioning of "your circulating pump," he wrote. "Look at it and see that it is running properly." He assumed most cars did not have a water-pressure gauge and offered a tip on how to determine if the pump was working. "Press any rubber connection in the water-circulating system between pump and cylinders to test by the pulsations there whether your pump is delivering properly or not."[92] Taking the pulse of the water pump was but one way early motorists, in their supervisory role, sought to keep their machines healthy by diagnosing early symptoms that might cause serious troubles.

Although water was the critical coolant, oil was the lifeblood of any engine or machine. Many motorists knew this from their prior experience with clocks, sewing machines, bicycles, and other devices with gears and other working metal parts, but the higher operating speeds of automobile engines and greater strains on running gear demanded one pay closer attention than ever to lubrication. Different parts of the automobile's mechanism also required different lubricants in varying frequencies, adding complexity to the owner's to-do list. Russell advised motorists to "have a look at [their] cylinder lubricators," after starting the engine, to "see if they are feeding [oil] properly."[93] In some automobiles, like the 1927 Packard, motorists had to operate a hand pump from the driver's seat while under way to oil the chassis "every day and keep it free from squeaks and rattles."[94] More important, operators had to monitor oil consumption, for cars of the era voraciously burned oil. A Ford Model N owner in 1906 used ten gallons of oil in going but five hundred miles, much of it wasted by being blown out through the crankcase breather tube and scattered over his engine, while in the 1910s and 1920s Ford Model Ts routinely burned a quart of oil every hundred miles when new, and more once the piston rings and other engine parts loosened with age. As late as 1934 new Studebakers normally went through a quart every two hundred miles.[95] Because engines throughout this period used so much oil, unlike today, when a well-maintained car might go thousands of miles without adding any, and because there was no instrument to tell owners whether they might need to add oil, motorists learned to frequently raise the hood to ritually check their oil.

It has always been a somewhat messy procedure, but on early automobiles checking the oil could be really dirty. Model T owners had to reach deep into the engine compartment to manipulate two little petcocks, or faucets, on the engine's crankcase. They first opened the lower petcock, near the bottom of the crankcase; if nothing came out, it meant the engine was almost dry and oil had to be added; if oil poured out, one quickly closed the petcock and opened the upper petcock near the top of the crankcase. If oil came out of this valve too, there was too much in the crankcase. The Ford Motor Company recommended keeping "the oil level at a point about midway between the two petcocks," reported an expert who pointedly added: "How this can be determined without the use of the X-ray can only be conjectured."[96] The Ford's arrangement made it almost impossible for owners to keep the oil supply at the factory-recommended level and was user unfriendly in the extreme: they either added too much or too little, wasting oil or risking injury to their engine. By the end of the 1920s, Ford finally switched to a dipstick for measuring the quantity of crankcase oil, the method already adopted by most other automakers. Withdrawing the dipstick, wiping it clean, and then reinserting and removing it again to read the oil level—as is still today the method for performing this check—did not take much time, but it was a chore motorists performed regularly and frequently.

As late as 1910, unless automobiles were electrics or steamers, they seldom came from the factory with any instruments. Electric cars always had gauges to inform operators of current discharge from their batteries, while steamers had gauges indicating crucial operating information like boiler pressure. The first instrument to be widely installed as standard equipment on gasoline cars was an oil pressure gauge.[97] Installed on the dashboard, this instrument made it possible for operators to tell at a glance if their engine's oil pump was functioning, eliminating the need to visually check the cylinder lubricators on the engine itself. If the gauge showed low oil pressure, it might indicate a low oil supply, although such instruments were only crude indicators of the quantity of oil in the engine. Although gasoline gauges were common by the early 1920s, only at the end of that decade did the development of new types of sensors allow the devices to move to the instrument panel, their present position, where operators could monitor their fuel supply without getting out of the car. Similar instruments emerged to permit motorists to monitor their engine's water temperature from the driver's seat. The first appeared in the 1910s, the Motometer, which was a popular aftermarket accessory, essentially a large thermometer that replaced an automobile's original radiator cap and that could be read through the windshield. A glance enabled operators to know if their engine's coolant was about to boil over. By the 1920s, dash-installed temperature gauges, conveying information using newly devised temperature sensors installed in the engine, supplanted the Motometer.

Those who got behind the wheel of the new 1912 Cadillac, the first car available with electric starting, noticed another new instrument on the dashboard: an ammeter. It monitored the car's electrical system, specifically the flow of current to and from the battery (and was present, obviously, on electric vehicles earlier). Owners were not always clear as to just what the ammeter's readings meant, and most operators probably learned that they had battery or generator troubles in some other way. Still, ammeters became standard equipment even on Model T Fords, at least those that, starting in 1919, had optional electric starting. The ammeter sat alone in the middle of the Ford's small instrument panel, centered on the wooden dash. Frugal buyers who bought their Fords without the costly option also got an instrument panel but with no instrument—just a metal blank where the ammeter might have been.

The most important instrument today, the speedometer, was long thought unnecessary and only became standard equipment in the 1920s. "To the casual observer," observed one early speedometer maker touting its product in a 1910 brochure, "an Auto-Meter is a speed indicator, an extravagance, a toy. To the thoughtful, careful driver it is a guardian." Today we might associate speedometers and the word *guardian* with people's safety, with the instrument's ability to remind operators as to how fast they were going, yet the manufacturer considered the Auto-Meter to be a guardian of the automobile and of the operator's enjoyment of it. The only use related to speed the company's brochure mentioned was helping operators "change gears without jerking the car." If through trial and error owners learned at what speed they could shift gears most smoothly, they could thereafter simply watch their Auto-Meters and shift at that precise speed.[98] They could also use their Auto-Meter to "check tire wear," comparing the number of miles a tire lasted with the mileage the tire maker guaranteed. The instrument could also prevent motorists from running out of gas: "You already know what distance a gallon of gasoline will carry you," explained the brochure, so "if you know how many gallons you start with, the Auto-Meter will tell you when the supply begins to run low." The ability of an Auto-Meter or any other speedometer to tell motorists how *far* they had traveled, which it did by counting miles—its odometer function—seems to have been more highly valued in the early decades of the auto age than its capacity to indicate how *fast* they were going. Motorists kept close track of mileage for a number of reasons in addition to estimating gas consumption. For one, early driving directions, which, in this era before road maps were available, gave instructions according to distance—start at the train station and head north on Railroad Avenue, and after 2.7 miles turn left on a dirt road at a stone wall, and so forth. Also, many owners kept logbooks, noting their daily mileage along with every penny spent on oil, gasoline, and repairs. At a time when debate raged as to whether keeping an automobile was less costly than keeping a horse and buggy, some motoring devotees worked out their per-mile operating costs to three decimal points.[99]

Then as now, most speedometers had two odometer registers, a secondary trip one, usually with three digits, along with a primary scale having five or six digits, often showing tenths of a mile as well. People used the trip odometer as we would today, resetting it before a trip or on filling up their gasoline tank. Unlike today's cars, however, where only the trip meter can be reset, many cars in the first two decades of the century allowed owners to reset *both* the trip as well as the primary odometer. This was due to the traditionally seasonal nature of motoring, where in all but the most temperate climates owners stored their vehicles in the late fall and took them out again only in the late spring. At that point, they would reset their primary odometers to zero for the new motoring season.[100] As year-round operation became the norm, and cars lasted longer and were likely to be resold as used vehicles, the mileage registered on their primary odometers came to be viewed as cumulative, the total distance driven over a car's lifetime. Setting an odometer back to zero came to be considered unethical and illegal.

As late as 1925, some Ford customers still viewed the optional speedometer on the Model T as a needless frill. "I ain't gonna be told I hafta" have a speedometer, one prospect told the dealer, having heard that such instruments would soon be standard equipment.[101] By the end of the decade, this had happened and car buyers no longer had any choice: all automobiles came with speedometers, along with gauges indicating oil pressure, water temperature, battery condition, and amount of gasoline in the tank as well. Although neither state nor federal legislation lay behind this increased instrumentation, surely the public attention given highway accidents, along with higher average speeds, had something to do with the speedometer's newfound importance. Its location also no longer was one of those "petty things" not requiring standardization, as an industry insider had argued back in 1917.[102] By the 1930s, the speedometer not only was integrated with other gauges into what commonly was now called the instrument panel, but it stood out as the largest and most central of those instruments. As it had in many areas, Chevrolet led the low-priced field in introducing these changes. Its 1928 cars clustered instruments around a speedometer on a panel, installed in the middle of the dashboard between the driver and passenger, an arrangement Ford immediately emulated on its new Model A.[103] For the next model year, Chevrolet moved the instrument panel directly in front of the driver, heralding what would become universal industry practice. By the 1934–35 model year, Chevrolet's speedometer was about 25 percent larger in diameter than the gauges surrounding it, while by 1936 it was 30 to 35 percent larger. In 1937, its diameter peaked at about eight inches, dwarfing the shrunken ancillary instruments that were clustered around it.[104] As if giant speedometers did not alone send a clear enough message, in 1938 Chevrolet incised the words Safety *First* into the chrome panel beneath the equally large but redesigned horizontal-format speedometer, which now stretched nearly a foot across the instrument panel. Although the Chevrolet was probably incapable of one hundred miles per hour, its speedometer was calibrated up to that speed, attesting to the industry's subtle privileging of performance over safety.[105]

By the latter 1930s, then, the instrument panel had found its modern form. Operators supervised their automobile's speed and monitored its condition and performance from the driver's seat, glancing at their dashboard gauges. Some cars already were appearing with what a 1934 reviewer called "telltale lights," an innovation that replaced ammeters and gauges registering oil pressure and water temperature. "If the oil pressure fails," the critic enthused, "a red light shines; if the generator is not charging, another red light shines; if the water is boiling, a red light glows; if the water in the battery is too low, a light indicates it; etc." He believed telltale, or indicator, lights—better known by many today as idiot lights—were labor saving. For the driver, he concluded, now "the only instrument that he need watch is the speedometer."[106]

By the mid-1930s, with or without telltale lights, motorists behind the wheel of a modern automobile had less to do than ever. No longer did they have to monitor the automobile's functioning so closely, or work so hard starting, steering, stopping, and shifting gears. Driving in 1935 no longer resembled the work of a factory hand tending a complex machine. Whereas operators in the first few decades of the century had to manually light their oil or gas lamps to run their cars at night, modern motorists simply flipped a switch or pulled a knob. Instead of hand cranking an engine to start their car, they depressed a button or turned a key. And while under way they no longer had to manually pump oil to lubricate the chassis or to pressurize the gasoline to force fuel to the carburetor; these and many other jobs that once had to be performed by operators were now automated. Inventions like synchromesh and hydraulic brakes lessened both the physical effort required to operate a car and relieved operators of some of the anxiety they had once felt in the driver's seat.

The word *operator*, in fact, no longer accurately described a person behind the wheel of a modern automobile. State motor vehicle departments would continue to use the term, but by the 1930s most people spoke instead of "drivers" or "motorists." Similarly, few people any longer referred to cars as machines, though the latest models were more mechanically complex than ever. Such shifts in usage recognized how dramatically the behind-the-wheel experience had changed over the previous two or three decades. For one thing, driving had become sedentary in ways unimaginable in the early years of motoring. Drivers now ran their cars almost exclusively from the comfort of the driver's seat. Even when they pulled into service stations to see after their vehicles' needs for fuel, oil, water or air, owners were often relieved by uniformed attendants of the need to get out from behind the wheel. Driving was not merely more sedentary but also more comfortable. As late as 1920, 90 percent of new automobiles were still open touring cars or runabouts, their operators unprotected from weather, road dust, noise, and the gaze of strangers. By the 1930s, the same percentage of new cars were coupes or sedans, offering far greater privacy and comfort. Taller or larger drivers had less trouble getting into the newer cars or tending to their controls. Whereas earlier vehicles had often required the skills of a contortionist—because of the steering column, one operator complained in 1913, "Instead of sitting straight, I had to sit sideways"—now driver's seats were universally adjustable, and on some makes so were the foot pedals and steering wheel position. By the end of the Depression decade the gearshift lever had migrated from its long dominant position on the floor in the center of the driver's compartment to the steering column, further relieving congestion in the front seat and adding to driver comfort.[107]

Freed of much labor and exertion, and comfortably ensconced in all-steel mobile cocoons, modern drivers had new options. Insulated from road noise, they could easily converse with their passengers, take in the scenery, or daydream as the miles smoothly rolled by beneath them. Or they could listen to the radio, an automotive accessory that recently had become hugely popular. Although radio technology, or "wireless," as it initially was known, predated the horseless carriage, it first became a consumer product in 1920, with the rise of radio broadcasting. During the twenties Americans flocked to stores to buy radio receivers, enthusiastically tuning in to live performances by orchestras like Paul Whiteman's, comedy shows like *Amos and Andy*, or sportscasters like Graham McNamee, announcer for the World Series. When commercial radios for automobiles first came on the market in 1929, the public understandably jumped at the chance to bring the new entertainment medium with them on the road. Sales of car radios grew rapidly, according to a historian of the technology, and "skyrocketed to 724,000 units" by 1933 and doubled again by 1935 in spite of the Depression.[108]

Not everybody was equally excited about the prospect of mixing motoring with listening, however, and critics strongly opposed the practice. "Better park the car before you turn on the radio," advised the author of a 1931 book on safety, while a writer from Consumers' Research, Inc., a private company then based in New York that tested and rated products, alleged that it was "almost criminal to permit the installation of equipment in an automobile which will divert the driver's attention and slow his reaction time."[109] Others called car radios "attention diverters" and urged their prohibition.[110] "The dulcet voice of a crooner may be a welcome passenger in an automobile which is traveling along on a straight and monotonous highway through the prairie states," conceded one critic, but under most conditions the car radio was a "distraction" and threat to safety.[111] Although these arguments anticipated those heard today against cell phone use while driving, no hard evidence was ever adduced as to the detrimental impact of car radios on safety. For their part, consumers blithely ignored the critics and motored on, accompanied by the beat of Duke Ellington or the crooning of Bing Crosby.

The success of car radio was another measure of how much easier it had become by the 1930s to run an automobile, a technology now more like an appliance than the confusing and unfamiliar medley of "dials, gages, levers, pedals and switches" confronted by early operators. While not quite as reliable and easy to use as a stove or refrigerator, a new automobile no longer was a machine demanding hard or stressful work; it also had become a proverbial living room on wheels, a site of comfortable leisure and enjoyment. Looking back from the Great Depression, a character in John Steinbeck's novel *East of Eden* summed up the transformation in people's relationship to this personal technology over forty years. "It is hard now to imagine the difficulty of learning to start, drive, and maintain an automobile," the character muses, recalling the early-twentieth-century experience. "Not only was the whole process complicated, but one had to start from scratch," that is, adults who grew up with the horse and buggy had to learn the unprecedented complexities of the horseless carriage. "Today's children breathe in the theory, habits, and idiosyncrasies of the internal combustion engine in their cradles," Steinbeck's character continued.[112] He might have added that, by the late 1930s, not only were children in cradles riding in automobiles but older Americans, too, had already been driving or at least riding in cars for a long time. The automobile had become an inseparable part of people's lives. Not everybody yet owned a car, but the technology had ceased to be exotic and strange. The Supreme Court of Minnesota made this point in a 1936 automobile negligence case. "No longer can it be said," wrote the judge in the majority opinion, "that jury men and women are inexperienced or inexpert in respect to the handling of an automobile. The ownership and driving of such vehicles are now so common that one may safely say that practically all jurors are experienced in respect of their operation."[113] In the language of the law, the judge had taken "judicial notice" of the ubiquity of that experience, meaning that no longer in such cases would plaintiffs have to call witnesses to testify as to how an automobile worked or what operators had to do to control the machine. It was another way of saying that a technology that was once radically new and intimidating had become, three or four decades later, traditional and taken for granted.

Contributing to the automobile's success were modern vehicles' much greater reliability. Earlier motorists had put up with machines that were far more fragile in construction and finicky in behavior; something was always going wrong, requiring adjustment, or demanding repair. Unless owners employed a chauffeur, they often had to use tools and tinker with their car to keep it running. When garages were few and auto mechanics scarce, motorists had to provide their own first line of technical support, especially when on a trip far from home. The enjoyment and exhilaration they experienced behind the wheel was inseparable from the greasy, hands-on, finger-pinching and knuckle-skinning work they frequently had to do under the hood. They had to earn whatever freedom their cars made possible. To those under-the-hood activities and a look at how owners and operators coped with car trouble we turn in the following chapter.

"Do you have one that can help me decide which is the one for me?"

Technology shopping has always been bewildering. Consumers must choose a particular machine, along with necessary accessories and options, without knowing much about what the new technology does. This 1980 cartoon makes fun of the first-time computer buyer who thinks the machine can make the choice for him. © *Frank Modell / The New Yorker Collection / www.cartoonbank.com*

ARRIVAL OF THE
No. 8 WHEELER & WILSON

On taking delivery of their new machine, consumers expect relief from drudgery, new powers, or wonderful entertainment. The "arrival" of the new Wheeler & Wilson sewing machine depicted on this 1885 trade card is an exciting moment, although the buyer has no idea what complexities and struggles await her as she learns to use the new technology. *Warshaw Collection of Business Americana—Sewing Machines, Archives Center, National Museum of American History, Smithsonian Institution*

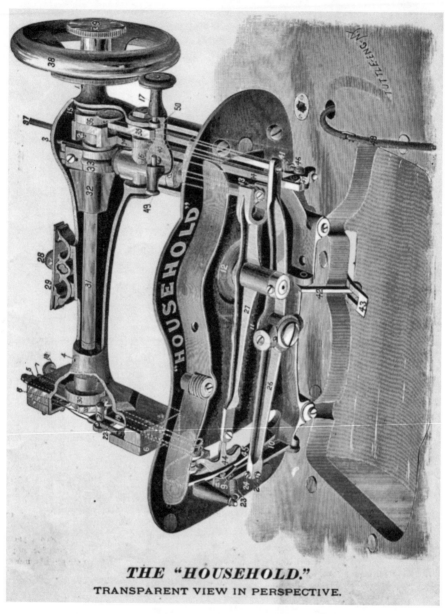

THE "HOUSEHOLD."
TRANSPARENT VIEW IN PERSPECTIVE.

Sewing machines came with the first owner's manuals, which described in considerable detail how to thread, manipulate, and maintain the device. Although most women already knew how to sew, doing so with a machine required new knowledge and skills. This illustration from a "Household" sewing machine owner's manual introduces the user to her new machine, depicting at once its exterior as well as an X-ray perspective on its innards. *Collection of author*

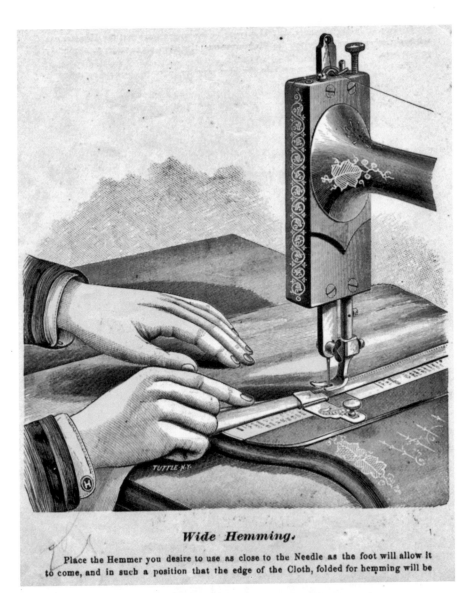

Wide Hemming.

Place the Hemmer you desire to use as close to the Needle as the foot will allow it to come, and in such a position that the edge of the Cloth, folded for hemming will be

This inventive hands-on illustration from the "Household" manual shows a woman "wide hemming" with the help of the machine. Such views helped the user to understand the task, to relate her hands and body to the machine, and perhaps to gain confidence in her ability to accomplish such work. *Collection of author*

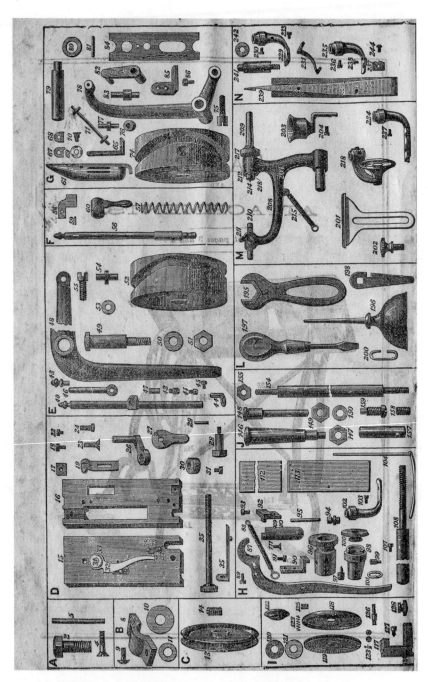

As one of the first technologies to enter the home, sewing machines introduced consumers to the challenges of maintaining and repairing household appliances. The 1871 Howe Sewing Machine manual was illustrated with numbered parts lists so that a user, even if she didn't know what something was called, could identify a

THESE NUMBERS AND NAMES REFER TO THE

PLATES OF PARTS

ON THE PRECEDING PAGES.

Order all Parts by the Numbers, and state what size Machine, A, B or C.

SECTION A.

2 Arm Screw.
4 Arm Plate Screw.
5 Arm Spool Pin.

SECTION B.

8 Shaft Cap.
9 Shaft Cap Screw.
10 Bed Plate Washer, ⅜×⅝.
11   do.   do.   ⅜×½.

SECTION C.

13 Pulley.
14 Pulley Screw.

SECTION D.

15 Outside Face Plate.
16 Inside Face Plate.
17 Presser Bar Guide Adjuster.
18 Presser Bar Guide Adjuster Screw.
19 Thread Controller.
22 Thread Controller Screw.
23 Inside Face Plate Screw.
24 Outside Face Plate Screw
25 Presser Screw.
26 Presser Slide.
27 Adjusting Cam
28 Adjusting Cam Screw.
30 Presser Screw Boss.
31 Presser Screw Boss Screw.
32 Take Up.
33 Take Up Spring.
34 Take Up Spring Screw
35 Take Up Adjuster.
37 Thread Guide Pin.
38 Thread Guide Pin (Drilled).
39 Thread Guide Pin Wire.

SECTION E.

40 Needle Bar.
41 Needle Bar Set Screw.
42 Needle Bar Thread Shield.

43 Needle Bar Thread Guide (Upper).
44 Needle Bar Thread Guide (Lower).
45 Needle Bar Thread Guide Screw.
46 Needle Bar Piston.
47 Needle Bar Piston Pin.
48 Needle Lever.
49 Needle Lever Stud.
50 Needle Lever Stud Washer.
51 Needle Lever Stud Nut.
52 Needle Cam.
53 Needle Cam Roll.
54 Needle Cam Roll Stud.
55 Needle Cam Screw.

SECTION F.

56 Presser Bar.
57 Presser Bar Spring.
58 Presser Bar Guide.
59 Presser Bar Guide Pin.
60 Lifter.

SECTION G.

61 Shuttle, complete.
65 Shuttle Latch.
67 Tension Plate (Upper).
68 Tension Plate (Lower).
69 Shuttle Tension Spring
70 Shuttle Tension Screw.
71 Bobbin.
74 Shuttle Cam.
75 Shuttle Cam Screw.
76 Shuttle Cam Roll.
77 Shuttle Cam Roll Stud.
78 Shuttle Lever.
79 Shuttle Lever Stud.
80 Shuttle Lever Stud Washer.
81 Shuttle Lever Stud Pin.
82 Shuttle Driver Link.
83 Shuttle Driver Stud.
84 Shuttle Driver Slide.
85 Shuttle Driver.
86 Shuttle Driver Screw.

needed part, order a replacement for her machine, and when it arrived install it using the wrench or screwdriver provided by the manufacturer. *Howe Sewing Machine Manual, "Plates of Parts" and Adjacent Key. Warshaw Collection of Business Americana—Sewing Machines, Archives Center, National Museum of American History, Smithsonian Institution*

The New AUTOCAR Control
Simple As a Pair of Reins

The control of Type XI Autocar brings automobile driving to the simplicity of horse driving. In the rim of the steering wheel, and forming parts of it, are set two grips, one at the right hand, one at the left. These two grips control the throttle and the spark, regulating the speed of the car from 3 to 35 miles an hour. This arrangement brings the steering and the speed regulating together, so that in all ordinary running the only position necessary for the hands is on the steering wheel.

To be sure this car has a gear shift lever and an emergency brake lever at the right of the driver. The gear shift lever, however, is needed only on particularly hard hills or very bad bits of road. The car loaded with four passengers will climb nearly all hills on the high gear. The foot brakes being ample for all ordinary use, the emergency brake is rarely required. Hence we say that except in extreme cases only one position is required of the hands in driving Type XI Autocar.

This car has four cylinder verticle motor of 16-20 horse power, double side-entrance tonneau, and the smartest lines of any car. It is extremely quiet and vibrationless in running while having great power in proportion to its weight. Type XI Autocar is the ideal car for the man who wants an up-to-date, powerful, four-passenger car without excessive weight. The price of Type XI is $2000. Catalogue giving full description of it and also of Type X Runabout $900, and Type VIII Tonneau $1400, together with name of dealer nearest you, sent free upon request.

THE AUTOCAR COMPANY, Ardmore, Pa.
Member Association Licensed Automobile Manufacturers.

This Autocar advertisement from 1905 claimed that the controls used to operate the machine were as "simple as a pair of reins," an argument aimed at convincing horse-age consumers they were up to getting behind the wheel of an automobile. *"The New AUTOCAR Control, Simple As a Pair of Reins," NW Ayer Advertising Agency Records, Archives Center, National Museum of American History, Smithsonian Institution*

This is where the salesman leaves you after you have bought the car. You are confronted with dials, gages, levers, pedals and switches used in the control of the car. But this book takes you under the car and into every part influenced by these various controls—all parts are laid bare—and you see the *reason* for everything. Remember,

A MINUTE OF CARE IS WORTH AN HOUR OF REPAIR

"This is where the salesman leaves you after you have bought the car," begins the caption on the frontispiece of *Everyman's Guide to Motor Efficiency*, a 1920 handbook sold to information-hungry car owners. Like many computer users later, early motorists were always looking for clearer and more useful printed advice and help in operating and maintaining their complex machines. *Collection of author*

Early motorists, as with other early adopters of new technologies, were their own first line of technical support in that any road trip might include car trouble. In this contemporary cartoon, an overly sympathetic wife seeks to console her husband, who is baffled by his machine's failure to run. *Unidentified period cartoon, reprinted in Jack Sophir Jr., "Get a Horse! A Unique Collection of Early Automotive Humor from "Turn of the Century" Magazines (Litchfield, IL: Laughin' Jack Good Humor Publications, 1989), unpaginated.*

The *helpmate*—Really, John, I'm puzzled, too!

Hub. Grease
every 500
miles.

Front
Spring
Hanger, Oil
every 200
miles

Front Spring
Hanger Bolt.
Oil every 200
miles

Steering Post
Bracket Grease
Cup Oil every
500 miles

Lubricate Engine and
Transmission by daily
replenishments through
breather tube. Oil
level in crank case
should be carried
slightly above
lower pet cock

Spindle Bolt.
Oil every 100
miles

Steering Ball
Socket. Oil every
100 miles

Commutator
Oil or Vaseline
every 200 miles

Fan Hub, Grease Cup
One complete turn
every 50 miles

Control Bracket
Oil every 400 miles

Steering gear
Internal Gear Case
Fill with grease every
5000 miles

Universal Joint, Grease
Cup. Fill with grease
every 300 miles

Drive Shaft Front
Bearing, Grease Cup.
Two complete turns
every 100 miles

Hub Brake
Cam. Oil
every 200
miles

Rear Spring
Hanger. Oil
every 200
miles

Rear Spring
Hanger. Oil
every 200
miles.

Differential. Fill with
Grease once every
600 miles

Where to Oil the Model T.
A drop or two of oil should occasionally be applied
to all small connections and joints
throughout the car.

Among the things owners needed to know and could learn from their instruction manuals was how to lubricate their cars. This Ford lubrication chart was typical in depicting the vehicle's bare chassis with annotations as to where oil or grease was needed and how frequently. Lubrication charts were all but universal in owner's manuals well into the 1920s. *Warshaw Collection of Business Americana—Automobile Industry, Archives Center, National Museum of American History, Smithsonian Institution*

This 1916 calendar illustration was captioned, "Knowledge is power," invoking an old adage that is doubly true for technology consumers. Although the image exploits an attractive female motorist to invite a presumably male gaze, the message that tools and texts were essential aids for early car owners is indisputable. Users still find themselves seeking advice, help, or information from manuals, self-help articles, or online FAQ pages. *Collection of author*

The PACKARD *Lubrication System*

"MRS. SMITH why don't you tell Mr. Smith that you will take care of the car if he will buy one? That you will lubricate it every day and keep it free from squeaks and rattles and at the same time, prevent the cost of wear and tear? Perhaps it is only fair to tell Mr. Smith that you could do this each morning in less time than it takes to sound the horn and with no more effort. Just a pull of this plunger and oil is on the way to the 38 chassis points requiring regular and daily attention.

A Packard is truly a woman's car when it comes to ease of care and this is getting to be more important every day with more and more women operating their own cars. Every car needs daily attention and the better the car, the more surely it ought to have it. What do you say, Mrs. Smith? Will you take the job if Mr. Smith buys the car?"

Many tasks that early motorists had to laboriously perform themselves, such as oiling chassis parts, were later automated, in full or in part. On the 1927 Packard an operator-activated plunger sent oil "on the way to the thirty-eight chassis points requiring regular daily attention," a user-friendly innovation that "Mrs. Smith" in this advertisement is happy to perform. *Warshaw Collection of Business Americana—Automobile Industry, Archives Center, National Museum of American History, Smithsonian Institution*

Car owners have always had to feed their machines, meeting their need for fluids and supplies. The women pictured here are adding water to a car's radiator. When this snapshot was taken in 1927, vehicles consumed far more water, oil, and air than today's cars, not to mention requiring more frequent replacement of tires, fan belts, spark plugs, and other components. Cars got about the same gas mileage as present vehicles, however. *Catalog no. 2006.0108-009, Division of Culture and the Arts, Photographic History Collection, National Museum of American History, Smithsonian Institution*

Considerable humor derives from our frustrations with technology, and the personal computer especially has given rise to many jokes and cartoons, like these fake error messages. Some make fun of the user's supposed ignorance, while others target the absurdity of the devices themselves and the bizarre illogic of their design. Laughing at technology troubles is one way we cope with the user unfriendly nature of these essential tools.

Chapter 4

Tools, Tinkering, and Trouble

I N 1908, A MISS McGAW, the daughter of a prominent Philadelphia businessman, bought a Wayne roadster. "I tinker around a good deal," she
told a local reporter, "just to satisfy my curiosity" about "the working parts
of my Wayne." Tinkering, she went on to explain, helped her learn "what
to do on the road" when trouble occurred.[1] Like other early automobilists,
she recognized an essential truth about the new technology: "Anyone of
average intelligence can learn to run a car in less than a day," as automotive writer Morris Hall put it, "but what to do on the road when anything
happens is the nightmare of the untutored driver."[2] Like many other early
motorists, Miss McGaw sought to tutor herself to avoid such nightmares.
Women especially might feel vulnerable being stranded alone, perhaps
far from civilization, with an inoperative machine; male automobilists
seemed more concerned about social ridicule if stranded through a breakdown. "There is nothing so embarrassing to an automobilist," one wrote,
"as to take a few friends out for a ride and then have something happen
which he is unable to straighten out."[3] Men felt pressured by the cultural
stereotype that equated masculinity with an ability to handle tools and
master machinery, yet most of the men who bought automobiles in the
first two decades of the twentieth century were from the business or professional classes and did not work with their hands, have experience with
tools, or possess much mechanical knowledge.

Early car buyers, then, like those who earlier had purchased clocks, sewing machines, or bicycles, also had to learn the parts and systems of their
machines and to use the tools that came with them to make adjustments
and repairs. For some people, the mechanical obligations of car ownership were grounds for not buying one; while for others like Miss McGaw

the automobile's technical complexity was something found enjoyable and rewarding. Christine Frederick, the well-known scientific management expert, celebrated the way "spark, throttle, cylinders, gear, magneto and steering wheel have yielded their secrets to me" and "wrought my emancipation, my freedom."[4] Studying the automobile's "construction and operation," enthused a male motorist, constituted "a liberal education in itself," so varied were the areas of knowledge and the intellectual challenges it provided.[5] Save for the relatively few car owners who engaged chauffeurs, some tinkering and involvement with their vehicles mechanisms was unavoidable as garages and service facilities were few and far apart in the early motor age. It was this frequent need to tinker, to replace or clean a spark plug, say, tighten or replace a fan belt, fuss with the igniter on the acetylene gas headlights, or remove a wheel to change a tire—not just the operational challenges of starting, steering, shifting, and stopping early automobiles—that made the technology so user unfriendly.

In this chapter, we look at car owners working "under the hood," a phrase I use to refer generally to all of the mechanical and operating components of a car, including brakes and tires, which by giving trouble invited motorists to get out, unpack their tools, and set to work.[6] The earliest automobiles did not even have hoods, of course, as their engines were mounted on the chassis amidship, beneath the driver's seat; operators gained access to them by lifting the seat, raising the entire body off the chassis, or crawling beneath the car, the latter indignity memorialized in the chorus of the 1913 popular song "He'd Have to Get Under—Get Out and Get Under (to Fix Up His Automobile)."[7] Already by 1905, however, engines had moved to their now familiar location in the front of the car, where they were covered with metal shrouds, or "hoods," an arrangement allowing for easier access to whatever was under the hood.

There was so much to know about these complicated machines, and the learning curve was steep. Owners acquired knowledge through a combination of reading, listening to others, and experience with machines. The automobile was commonly analogized to a sentient, living creature subject to ills and maladies that an alert owner-user might detect through careful observation. While in the nineteenth century, users sometimes attributed "spells" or "fits" to their sewing machines, automobilists pushed the metaphor of illness much further. In its title, *Dyke's*

Diseases of a Gasoline Automobile and How to Cure Them, published in 1903 and one of the first such handbooks to appear in the United States, exemplified the tendency to analogize cars to people.[8] "The automobile, like the human body," observed another author, "may become deranged and require a remedy."[9] Similarly, a Ford Motor Company brochure explained that even "the best of motor cars sometimes need a physician's attention" and sought to reassure prospective buyers that "every Ford agent is a Ford Physician."[10] Other metaphors described "sick" cars going to the "hospital," towed there by an "auto ambulance," and women studying auto mechanics in a YMCA class as starting "to dissect one of the cars as a doctor does to a cadaver."[11] People routinely referred to the "anatomy" of automobiles: the carburetor was the "lungs" through which the engine "breathes,"[12] while the engine became the "vital beating heart" of the machine.[13] Oil "is its blood," noted another, as its engine "will not continue to run long without the proper supply of lubricating oil, and when it stops the result will be akin to death in the case of a human being."[14] And long before onboard computers provided automobiles with anything like an ability to think or process information, some attributed sentience to automobiles, like the editor who confessed that sometimes it was "pretty hard to tell what is going on in the mind of any motor car."[15]

The metaphor everybody borrowed from medicine and used when speaking of car trouble was "diagnosis." This reflected the authority and prestige of early-twentieth-century medicine but also was an apt analogy: both motorist and doctor relied on detailed, difficult knowledge and theory and both closely observed their subjects, looking for symptoms, arriving at a hypothesis, and then confirming or rejecting it by testing and examination. Car owners hoped that correct diagnosis might lead to effective treatment or even a cure for what ailed their car. Every car owner wanted "to know how to find and overcome the ordinary ills and troubles" of their machines, as a handbook author wrote in 1918, "and how to diagnose and prescribe for it when it begins to wheeze or squeak or groan or knock."[16] Yet not only car owners but even professional mechanics could be baffled by conflicting or false symptoms and, no less than a physician striving to determine the origin of a patient's distress, struggle to reach a diagnosis. While early-twentieth-century doctors benefited from various diagnostic technologies such as X-rays, cardiographs, sphygmomanometers, and blood tests, motorists and mechanics had only their eyes, ears, noses, and hands to rely on.

What we might call "learning by listening" became a newly important source of knowledge to auto-age consumers. "When a working part of an automobile ceases to perform its function perfectly," a 1907 handbook taught readers, "it gives prompt notice of that fact in language almost as plain as articulate words"—a reference to the noises cars make.[17] Another article, "How a Gas Engine Talks," advised motorists that when an engine was running right "the only sound it emits are [those] such as are made up from the clicking of the valves, the inhalation of the air and the exhaust." Sick engines "spoke" very differently: "A loose flywheel causes a thump, usually at every impulse the engine takes. Looseness at the wrist box [where the piston connects to the rod] causes a knock. A loose crosshead box [the bearing linking connecting rod to crankshaft] usually causes a clatter."[18] Despite confident assertions that owners could "know by the sounds which issue from the engine and the gearing what is happening in the dark recesses—whether the monster has indigestion, whether he is coughing, swearing, or only cooing to himself," this knowledge was difficult to acquire.[19] Learning to distinguish a "wheeze or squeak or groan or knock," as another author put it, took time.[20]

One sound the greenest of novices could recognize was silence: when their engine suddenly stopped or refused to start. To diagnose what was wrong, they needed a rudimentary understanding of how a gas engine worked. Motorists needed to know that their engine's carburetor mixed gasoline with air and that this mixture, under pressure in the cylinder, burned when the electrical system introduced a spark. Possessed of such knowledge, they were urged to systematically eliminate one possible cause after another. They first looked to see if they had gas in their tank. If the fuel supply checked out, then they made sure gas was getting to the carburetor (either fed by gravity or under pressure). Motorists might discover a broken or leaking fuel line was the cause of their difficulty. If gas was getting to the carburetor, possibly its jet or filter was clogged with grit.[21] If gasoline was reaching the engine, however, the trouble probably lay with the electrical system. Few motorists understood much about electricity, but many learned enough about their vehicles to check some basics: to look for a loose or detached battery cable or wire running to the engine's spark plugs, as these could prevent ignition. Spark plugs themselves were fragile in the early days, as their porcelain insulators often cracked, allowing current to leak or dissipate, which would prevent ignition, while poor combustion frequently fouled the points of plugs with carbon, producing the same result. Owners learned to remove a suspect plug, check its cleanliness, and if necessary reset the "gap" between its two electrodes. Many also knew how to test the battery to determine its charge. Motorists might get help in diagnosing trouble from their owner's manuals, which often outlined simple tests or examinations to determine if the battery were charged or if a spark plug had failed. Some manuals cleverly represented such knowledge in step-by-step diagnostic charts, breaking down the larger diagnostic problem—why doesn't the engine start, or why did it quit?—into discrete smaller questions owners could better answer.[22]

Most tinkering with automobiles required tools, and first-time owners often had to familiarize themselves with those packed in their car's tool kit. The number and kind of tools that came with new vehicles varied but was far greater than what purchasers of sewing machines or bicycles received. Buyers of expensive cars tended to get a more comprehensive set of tools than did purchasers of medium- or low-priced vehicles. Buyers of a 1923 Dodge, a popular mid-priced car selling under a thousand dollars, received a grease gun; three socket wrenches; various tire and wheel tools, including a hubcap wrench, tire pump, wheel puller, and demountable rim wrench; a valve-grinding tool; a center punch; a ball-peen hammer; various screwdrivers; and a selection of open-end and specialty wrenches (fitting the horn, ignition unit, etc.).[23] Purchasers of the 1919 air-cooled Franklin, which cost upward of twenty-four hundred dollars, received not only the tools standard with the Dodge but also a greater variety of wrenches along with a selection of spare parts, such as fuses, bulbs, and spark plugs.[24] Those affluent few who could afford to pay three times the Franklin's cost for a Pierce-Arrow acquired an even more extensive inventory of tools and parts. The car came with numerous open-end wrenches, different-sized monkey wrenches, and more than half a dozen specialized wrenches that fit the carburetor, distributor, headlights, brakes, and demountable wheel rims, plus an inspection lamp that ran off the battery.[25]

That owners were expected to learn the use of their tools is underlined by the fact that automakers advertised the ease with which operators could access them in their vehicles. "Tools are not frequently used on Packards," bragged a 1916 promotional brochure, "but when one wants tools it is comforting to find them in the most convenient place—the door pocket"—a reference to the custom-fitted storage places for tools found on expensive cars, either inside the vehicle or on its running boards, permitting access to tools at a moment's notice.[26] Buyers of less expensive automobiles found their tools bundled in a leather or cloth tool "roll," shoved under the front or rear seat. Because tools were easily misplaced or left by the roadside, or because owners thought more or better tools might make tinkering easier, a sizeable tools aftermarket grew up around the automobile. In their 1911 catalog, the Motor Car Equipment Company, one of many tool suppliers, offered an "ideal owner's outfit" with thirty-four tools for sixteen dollars, plus a more basic twenty-piece "Tourist outfit" for $8.50.[27] In 1916, another firm's catalog listed its "No. 1 Krackerjack Tool Kit" with thirty-nine tools for fifteen dollars and a "No. 3 Krackerjack Runabout" version containing twenty-nine tools for six dollars.[28] Emphasizing the numbers of tools in each set, catalogs suggested that the more equipment one owned, the better prepared one was to take on any challenge, somewhat like Daniel Defoe's character of Robinson Crusoe who, following a shipwreck, collected a vast trove of miscellaneous items, enabling him to survive while his shipmates perished.

Although most automotive hand tools did not require great skill, neophytes nevertheless struggled. Owner's manuals never instructed users how to hold a tool or how to effectively manipulate it. Such knowledge is often called tacit, meaning it is knowledge that is difficult or even impossible to convey in words, and so car buyers were left to their own devices in learning to use the tools that came with their cars or that they later acquired. Some people clearly had more of a knack for tool use than others, as Winifred Dixon's 1921 account of her cross-country road trip in a Cadillac Eight with her friend and the car's co-owner, Katherine, or "Toby," suggests. Dixon was the "diagnostician," as she put it, while her friend Toby wielded the tools, a division of labor that worked for them. "It was evident the brakes must be tightened if we were to reach the bottom of the canyon alive," Dixon wrote in describing one example of such collaboration. She got out, crawled beneath the car, and diagnosed the problem as a loose nut "which looked as if it connected with the brake." Crawling out from under the vehicle, she enlisted Toby, "who is exceedingly clever with tools and something of a contortionist" to return to under the car, wrench in hand, where she "managed to tighten" the wayward nut. The brakes held, and the two women successfully reached the canyon floor. "Never had we known a prouder moment," Dixon concluded, extolling their teamwork, and "the incident gave us courage to meet new contingencies and never again did I experience just that sick feeling of helplessness."[29]

WHEREVER MOTORISTS TRAVELED in those pioneering days, they were likely to suffer mechanical failures, or "accidents," to use their word. "Only one accident having occurred thus far—a broken steering knuckle," wrote a 1912 traveler.[30] "One tire puncture was our only accident," reported another of his 1922 trip from Chicago to Los Angeles.[31] He was lucky because, as an automobile handbook explained, "There is probably no machine in use which operates under such adverse conditions." The automobile "is exposed to dust, mud, and rain; and although the more delicate parts are enclosed, there are still many elements that require constant cleaning and attention."[32] Trouble was part of the language of motoring: springs and axles "broke," gears were "stripped," connecting rods were "thrown," head gaskets "blew," tires suffered "blowouts," engines "seized up," and transmissions "dropped." As the terminology describing these accidents suggests, motorists considered any such failure as violent even when they themselves were unharmed; any such breakdown left the machine inoperable or partly destroyed. Operators either had to make an emergency repair to enable their vehicle to limp home or they had to "get a horse," submitting to the ignominy of being towed by an obliging farmer.[33]

The subject of accidents and "first aid on the road" for cars was much discussed in books and magazines, which regularly published columns such as "Lessons from the Road" and "My Best Repair."[34] Early operators reported many ingenious fixes for roadside emergencies. On a trip to Colorado with her mother in 1907, for example, one Vera Marie Teape had considerable trouble because the two drive chains that transmitted power from her engine to the car's rear wheels kept jumping off their sprocket wheels. The two women had to stop frequently to wrestle the chain back onto one or the other of its sprocket wheels. When one of the chains finally broke, Teape stitched it together again using copper wire. When that repair failed, she reconnected the chain first with a bent steel safety pin and after another failure with "small rope twine." All these fixes, however, had "the same futile result" as the chain quickly parted. "I remembered," Teape recalled, that "someone had once told us that if we brake a chain we could bind the small sprocket wheel with the broken chain and fasten it stationary to some part of the machine and go on one chain." Doing this "the machine worked as well as ever," allowing Teape and her mother to drive to the nearest town, where they had the broken link repaired.[35]

If Teape drew on needlework experience in her attempts to stitch the chain back together, some motorists turned to Daniel Boone–like woodcraft in fashioning emergency repairs, such as the man who cut down a tree along the roadside to use in repairing his car. He had skidded into a ditch, turned his car over, and in the process "dished," or collapsed, the wooden spokes of one of his rear wheels. He righted the car, which was operable but for the broken wheel. Like other early motorists, he carried a spare *tire* but not, as we do today, a tire already mounted on a spare wheel. So what to do? As he still had three good wheels, he thought that if the broken, fourth wheel somehow could be kept from dragging on the ground, he could still drive the car. Getting out the ax he carried in his tool kit, he cut down a medium-sized tree and lashed one end of it to the broken wheel and the other to the front axle. By redistributing his passengers and other movable items so as to place as much weight as possible over the three good wheels, he in effect transformed his car into a more or less stable tricycle. Now he could slowly drive it, the three good wheels rolling on the ground while the collapsed fourth wheel hovered in midair, prevented from hitting the road by the rigid tree trunk. Heroic improvisations in the face of accidents provided good yarns to share with other automobilists who might wish to add ideas to their own repertoire of possible emergency repairs.[36]

Most of the first aid motorists administered to their machines, however, was more prosaic. The continuous bouncing of vehicles over the rutted paths that passed for roads in the first two decades of the century created more everyday sorts of accidents, including the leaks of vital fluids: oil, gasoline, water, and air. Losing air from pneumatic tires, and tire trouble generally, as one manufacturer's publication put it, "like the poor is always with us,"[37] Manufactured of natural rubber layered with cloth, early tires were easily cut or punctured and damaged by petroleum products, sunlight, and water, all things common to an automobile's operating environment. That tires also were usually inflated to relatively high pressures, sixty to eighty pounds or so, added to their propensity to leak or explode.

Changing a tire has never been easy, but before World War I it was a major ordeal. In those years motorists traveled with one or more spare tires stowed atop their rear trunk or on their front fenders, protected from sunlight and water by canvas or metal covers. But because spare tires were not already mounted on spare wheels, early motorists faced a much more arduous and complicated process when making a repair. First they jacked up their car so as to be able to remove the wheel with the damaged tire. This itself was often a challenge. "Jacking is easy in theory, or in a garage," explained Winifred Dixon after her transcontinental trip in 1920, "but the trouble with the outdoor art is that the car usually lands in a position where it has to be jacked up in order to get planks under it in order to jack it up."[38] She was referring to the situation, common before the coming of paved roads and shoulders, where the ground under the car was so yielding that jacking up the heavy car resulted instead in driving the jack down into the earth. Once the car was raised and the problematic wheel and tire removed, the motorist had to get the tire off the wheel, a physically demanding effort requiring the use of tire irons and even a second set of hands. With the tire off, the motorist extracted the inner tube from within it and searched for the leak. Patching a leak in the tube was itself a complicated process that ended with motorists having to "vulcanize" a soft rubber patch by heating it with a kerosene burner or other source of heat, such as the fuel modules included with the Shales Vulcanizing Kit, which, when clamped over the patch and ignited, heated the rubber to the required temperature in but "five minutes."[39] Once the tube was repaired, however, the motorist would then reinsert it into either the old tire if still serviceable or a new one and then force it back onto the wheel, a task that again required the use of tire tools and considerable strength. Finally, the wheel with its repaired tire would be reattached to the car. With the wheel reinstalled, the motorist completed the tire change by inflating the tire to the appropriate pressure, also grueling work, with only a hand pump. At that point the motorist could lower the car with the jack and prepare to get under way.

Oil, gasoline, and water leaks were less onerous than flat tires to deal with but also required immediate attention. Motorists often lost all the oil from their engines when the drain plug, located at the bottom of the engine crankcase, vibrated loose or was knocked off by stones on the high-crowned roads of the day. Without oil, an engine would quickly seize up and self-destruct. If operators discovered a missing crankcase plug in time, they might carve a substitute from a stick or stuff wadded-up cloth into the drain hole so that they might refill the crankcase (most carried extra oil) and continue their travels. Slow oil leaks, such as those caused by worn or faulty seals, could be discovered and monitored by owners, as today, by looking for telltale puddles beneath the car or in the engine compartment. Gasoline leaks were also very common. "As most drivers know," a writer in *Horseless Age* explained in 1908, "the copper fuel pipes have an unhappy habit of breaking at unexpected times, the favorite places being close to unions," that is, where the ends of such pipes were flared and held tight by threaded or welded collars to fuel pumps, gas tanks, and carburetors.[40] It was at such junctions that the constant engine vibrations and vehicle jarring caused the tubing to develop cracks and fail. That many early fuel systems were pressurized further inclined such pipes to leak. In the open cars of the day, motorists were less likely to discover gasoline leaks with their noses, so they relied on visual inspection to discover such problems.

All automobile owners had to worry about leaks or loss of water, but only with steamers and gasoline cars were such losses likely to produce a roadside emergency. With electric vehicles, the only water was in the batteries, and owners had to be sure the electrolyte level in each cell was correct and periodically tested for acidity. With steamers and gasoline cars, however, water losses could lead to serious damage to the machine. Steam cars literally ran on water, their engines converting it to steam, which then was exhausted into the air. Even when they had no leaks, steamer operators frequently had to refill their water tanks, often after only twenty-five miles or so of driving. Most gasoline cars used water to cool their engines, a system that theoretically should not have consumed water but which in practice always did. Water escaped from leaky hoses and connections, while in warm weather, engines overheated and "boiled over," causing steam and water to explode out of the radiator. To replenish their supply, motorists tended to carry a pail or other container for fetching water when on the road.

Automobile radiators themselves frequently leaked. "Making the water-containing parts strong enough to stand the vibration of service while having walls so thin as to offer the least possible resistance to the passage of heat from the water to the air" constituted the challenge of designing and manufacturing such components.[41] The soldered joints connecting delicate water-carrying copper tubes vibrated loose, while the tubes themselves were often punctured by stones thrown up from the road. And well into the 1920s Ford, which had devoted more effort and expense than any manufacturer to designing and mass-producing radiators, still was plagued by new units that leaked.[42]

To stanch leaks of vital liquids from their machines, early motorists tried many tricks. Sometimes simple tinkering, tightening a screw or nut on a fitting or wrapping a leaky water hose or gasoline line with sticky "tire tape" or ordinary twine, did the job. One motorist recommended wrapping a cracked copper tube that carried gasoline with "a portion of an old kid glove, treated with shellac." Admitting that "very few motorists carry shellac," he reported that one owner achieved a "temporary repair" by applying "one of the 'liquid court plasters' placed upon a piece of cleaning rag." He was referring to an eighteenth-century trick to hide lesions or blemishes on women's faces, hence the name "court plasters," which were made of an adhesive-like substance, usually isinglass or gelatin, covered by a piece of silk. Yet the man who employed a court plaster to stop a gasoline leak was "unwilling to trust this repair longer than was necessary to reach a copper shop," that is, a metalsmith, to get a more permanent fix.[43] Motorists showed perhaps the greatest creativity in patching leaky radiators. While proper radiator repairs required removing the unit from the car and taking it to a metalworking specialist who could locate the leaking tubes and solder them shut, Americans improvised and attempted to stop leaks using flax seed, cornmeal, bran, and even rye bread. A 1909 handbook described the technique. "A slight leak in a radiator can be most efficiently, although temporarily," stanched by means of "paste made from bread kneaded with the finger." The paste should "be well kneaded, then spread over the leaky part, and worked in with some tool which will do duty as a spatula, just in the way painters work up their colors on a palette."[44] Anything that dried solid with heat and then would not dissolve to easily in water might temporarily staunch leaks caused by cracks or small holes in water circulating tubes and be an effective Band-Aid for a leaky radiator.

Using familiar materials like cornmeal, bread, safety pins, or other items found in barns, kitchens, woodlots, and medicine cabinets to repair complex precision machinery represented a kind of folk practice, cutting against the grain of behavior recommended by automakers and experts who opposed such primitive remedies. Victor Pagé, an independent engineer and widely read automotive authority, warned against patching radiators with homemade or even commercial pastes. "Compounds of this nature," he wrote, "should never be used in a radiator that can be repaired by any other means," simply because they could block narrow water passages and render the radiator worthless. Such fixes might be justifiable as "a desperate last resort," Pagé conceded, yet such "little dodges and makeshifts," as another critic termed them, ultimately harmed machines. Yet they also were of a piece with an older tradition of creative automotive bricolage whereby motorists employed sheets of paper or playing cards to make temporary gaskets; hairpins to replace missing cotter pins in throttle linkages; "finely sifted wood ashes and lemon juice" as a metal polish; or chewing gum, florist's, or bailing wire as a substitute for welds or fasteners to hold components together under the hood.[45]

BESIDES HAVING TO TINKER WITH THEIR CAR when it broke down along the road, early motorists performed much routine maintenance. As noted, owners could not just hop in and drive off as we do today, or if they did so they greatly increased their chance of trouble. "To keep a machine in a state of perfection," a physician suggested in 1905, "one should devote every morning from ten to forty-five minutes to carefully oiling and looking over different parts, especially looking over tires, so as to be sure of not being detained" while under way.[46] In addition to providing the frequent lubrication early vehicles required and oiling the components needing it, owners had to manually "turn down" the grease cups located at strategic locations such as wheel bearings, suspension joints, and axles; the work wasn't hard but required a wrench and a few minutes' time. "If you spend twenty to thirty minutes once a week," a 1924 Nash owner's manual advised, "your chassis will be kept thoroughly lubricated."[47] Owners also checked their oil and water supplies, often before starting out in their cars and periodically while on the road. The fragility of early vehicles encouraged owner attention. "I have to work on the car for a full hour every time I wish to take it out," complained a Pittsburgh owner of his one-year-old car in 1911.[48] The absence of any "locking devices for the nuts or bolts" on his machine explained some of his work, as he had to tighten loose fasteners before every trip. Even so, he reported "losing more or less of my car ever since." That the streets of the era were strewn with car parts attests to the fact that "losing" one's car this way was not uncommon. In 1918, journalist Clifford Brokaw "counted twenty-six different species of other lost parts in crossing Fifty-seventh Street at Eighth Avenue, New York City."[49] Having loose or poorly connected parts was not just annoying and potentially dangerous but also a cause of what the Pittsburgh man called "too much noise." Owners often took the squeaks and rattles made by their car personally, as indictments of their choice of make or of their competency as an automobilist.

As a result, many car owners picked up their tools and tinkered in a "search for silence," spending considerable time hunting down the source of noises that some considered a source of "mental strain."[50] Because of the way early automobiles were constructed, they created a veritable orchestra of unintended sounds as the vehicles aged. When wooden-spoke wheels dried out, they squeaked, creaked, or groaned, a condition motorists alleviated by periodically turning a water hose on them, hoping that the spokes would swell and no longer move in the felloe, or the segment of the rim providing their support, and cause noise.[51] Metal body panels also rubbed noisily against each other as the wooden frames to which they were fastened dried out, resulting in bodies that flexed more than when they were new. "To prevent squeaks," a 1916 Buick owner's manual added, "the bolts which fasten the body to the frame should be kept tight."[52] The loudest part of the car, the brass section of the cacophonous orchestra so to speak, was the engine. Although even a well-running motor could be loud, all kinds of abnormalities introduced new noises. One annoying sound came from the leather belts transferring power from the engine to generators and water pumps. When the leather stretched and began to slip on its pulleys, the belts could emit a piercing, banshee-like shriek. Worn bearings, pistons, gaskets, valves, and other engine components provided sharp staccato knocks, poundings, or other percussive noises, while a leaking or "blown" head or carburetor gasket could produce loud whistling sounds. As a percussive counterpoint to all these noises, the steel rods and cables connecting steering gear, brakes, and carburetor to their respective controls all rattled and clanked beneath the car like some crazed drum section. Noises from beneath the vehicle became especially pronounced after the factory-installed rubber or fiber bushings, which originally reduced noise by damping the vibration of whatever control cables or rods ran through them, dried up and disintegrated.

When open touring cars and roadsters gave way to closed sedans and coupes in the 1920s, the search for silence accelerated. Riding in a comfortable environment separated by steel and glass from external road noises, motorists heard unwanted sounds more clearly and found them that much more annoying. Even Ford buyers, often inured to the racket produced by the rattle-prone Model T, now complained about noises in the new Model A. "I have found it a wonderful automobile for the price and am well satisfied with it in every way," one purchaser wrote Ford about his 1928 Model A Ford—"except there is a roaring noise inside the enclosed cars." He had heard other owners complain about the same sound, which he believed came from a sheet of metal under the floor in the rear of the car. "This roaring is somewhat more pronounced in your car than in the Chevrolet or Whippet," he concluded, making the kind of astute brand-to-brand comparisons common among technology consumers.[53] Another man reported that his brother canceled his order for a Model A "on account of this roaring noise." Still another purchaser wrote from Saginaw, Michigan, to complain about his 1928 Tudor Sedan. "I have purchased a new Ford car each year for the past eight years," he said, and acknowledged that "in an entirely new product there is bound to be some defects." Still, he felt "some constructive criticism" was in order and urged Ford "to change the method of ventilation" in the car as the windshield could not be opened at speed over thirty miles an hour "on account of the whistling and howling, which are so terrific that they become unbearable."[54] Even if an automobile owner diagnosed the source of such noises, the vehicle would often need to be redesigned to silence them, something beyond the reach of ordinary tinkering.

A few owners wielded tools in an effort to change their cars, to improve performance, or to alter function. Sometimes, this tinkering was minimal, as with Model T owners who, frustrated by their car's propensity to overheat, bent the blades of their engine's fan to a greater angle, thinking the change would draw more air through the radiator and thus provide better cooling. The modification made matters worse, according to Ford engineers, who noted how the engine would have to work harder to turn the altered fan and would therefore run hotter.[55] Other owners tinkered more radically, producing the precursors of what later would be called "hot rods" or "custom" cars. Particularly did the Model T Ford lend itself to such projects, for the car came from the factory with considerable power, and by the mid-1910s there were so many around that buyers could buy a used flivver for little money. Many efforts to get more power out of Model Ts, however, fell prey to the same kind of technical ignorance afflicting those who bent their radiator fan blades. This was the case with those who decided that if one spark plug per cylinder was good, two would be better. "It is frequently thought by those who have not given the subject much consideration," wrote an engineering authority responding to such efforts, "that to screw an extra plug into the cylinder at any place convenient will produce an increase in power." Yet "few engines are readily adaptable for an extra plug without making alterations to the cylinder," he explained, "and very often these entail considerable risk of injury" to the engine.[56] By the 1910s, those desiring hotter performance from Fords could, as historian David Lucsko has recently shown, buy ready-made performance components for their cars. These included special manifolds and exhaust systems as well as kits to convert the stock engine to a more powerful overhead valve unit. Similarly, Model T owners could turn to the aftermarket and purchase new bodies to fit their chassis, transforming their Tin Lizzie into a sleek sedan, torpedo racer, tractor, truck, or snowmobile.[57] Henry Ford opposed such conversions and only belatedly entered tractor and truck production, but energetic tinkerers could not be stopped.

Probably the most common tinkering aside from maintenance that early owners engaged in was accessorizing their cars, that is, adding mirrors, lights, additional instruments, anti-rattle devices, Moto-meters, trunks, luggage racks, and so on. Cars were minimally equipped as they came from the factory, so buyers were obliged to complete their machine according to their own need or to distinguish it from the thousands of other identical vehicles. It was easy to attach additional equipment to cars that, beneath their sheet metal skins, had wood frames, as one could use traditional screws or nuts and bolts. This accessorizing declined in the 1930s, when, as historian Kathy Franz has observed, all-steel body construction arrived, necessitating special tools and metalworking skills in order to readily attach things to automobiles.

Some tinkering was motivated by an experimental mentality, as owners tried out different equipment or fuels, additives, and coolants, thinking they might come up with something better than the manufacturer. "I have not been able to resist the temptation to tinker with my automobile and try various experiments in hope of improving it," confessed one early motorist.[58] He and others exchanged spark plugs, carburetors, and other easily detached components of their cars to see if another version might produce greater speed or better mileage. "Some of the people here are using kerosene to cool their motors in the winter instead of water," reported a car owner from Mankato, Minnesota, who asked what the editor of *Automobile* thought of the practice: "highly inadvisable," replied the editor. Gullible motorists also experimented with various gas-saving or combustion-enhancing devices such as the "Helix Gas Mixer," claimed by its inventor to be the only carburetor that "really carburetes." It was said that the Helix would "increase an engine's power, improve fuel economy, prevent carbon build-up, reduce shifting, and provide quicker acceleration, higher speed, quieter running and no sickening odor from exhaust."[59] While most automotive experts considered such devices worthless, they nevertheless encouraged experimentation by motorists. In a 1917 article titled "Accessories of Value," respected automotive writer Morris Hall suggested car owners might benefit from replacing certain equipment on their vehicles. Buying a larger gas tank, for example, would enable motorists to "buy fuel in larger quantities and thus get lower prices," he argued, and by filling up less frequently they might also minimize "loss by spilling and leaking." A "supplementary carburetor," he also suggested, might improve fuel economy: it would allow "running more slowly, which saves gasoline and faster [sic] on less fuel, saving thus at both ends, and at the middle as well," a claim that if it made any sense at all also defied physics.[60]

Although motorists knew next to nothing about the relatively new hydrocarbons categorized as "gasoline," they also tinkered with the fuel they used in their engines. Many believed that adding a few mothballs to each tankful of gasoline boosted power and eased starting.[61] A motorist who observed another owner filling his fuel tank with an unusual concoction wrote to *Fordowner* inquiring as to the efficacy of such mixtures. He reported that the man "got five gallons of gasoline, one gallon of kerosene, and then cut a piece of camphor, about 2/3–inch square, and dropped it into the tank" along with the liquids. The editor replied that "the idea" of mixing camphor with gasoline was "one of those motoring superstitions that has lurked among car owners for many years." It was "the general opinion, of automobile engineers," he added, "that it is better not to use camphor in the gasoline" as it gummed up the innards of the engine.[62] During World War One, fuel shortages prompted refiners to dilute gasoline with related but less volatile hydrocarbons such as kerosene, and many motorists complained that this resulted in harder starting and more carbon buildup in their engines. Nevertheless, the idea spread that one might run a car wholly on kerosene—or on benzene or naphtha, and some experimented along such lines. Although internal combustion engines can run on any of these as well as a variety of other fuels, engineers consistently pointed out that for optimal performance engines, and especially their carburetors, had to be specifically designed for such alternative fuels. Scientific and engineering evidence to the contrary, the popular belief in the miraculous potential of gasoline additives, alternative fuels, or miracle mileage boosters has long been the automotive equivalent of the nineteenth-century American craze for patent medicines, or that for health supplements in our own time. None of these substitute fuels or additives significantly improved performance, and most added to harmful carbon buildup.

Carbon in the cylinders of gasoline engines caused preignition, that is, the premature burning of the gasoline-air mixture, which resulted in a "peculiar metallic or ringing knock in the cylinder" along with a loss of power, especially while climbing hills.[63] Carbon resulted not simply from substances other than gasoline in an automobile's fuel but also from incomplete combustion, which was all but unavoidable in early gas engines. "It is not unusual for one to hear an automobile driver complain that the car he drives is not as responsive as it was when new after he has run it but very few months," noted author Victor Pagé. This might occur even when "there does not seem to be anything actually wrong with the car." The trouble, he explained, "is often due to nothing more serious than accumulations of carbon."[64] The substance was "one of the most bothersome of the motorist's trials," observed another writer. "Unfortunately," he continued, "no carburetor is at present made which gives a perfectly combustible mixture, nor is any engine built so perfectly that excess oil will not leak past the rings," which when burned also left carbon as its by-product. "Carbon," he concluded, "is forming continually, and, sooner or later, will have to be removed."[65]

People tried all sorts of physical, chemical, and technical means to eliminate this nasty and harmful substance. Some advocated "burning it out," that is, opening up the engine and using a blowtorch. This was dangerous and required equipment that most owners lacked, however, so many tried commercial remedies—products like Michener's chain carbon remover, sold by automobile supply houses. Consisting of fine brass or copper chain, owners would remove the spark plugs from their engine, drop a length of chain into each cylinder, replace the plugs, and then start their engine. The chain "is thrown around by the piston so as to knock off the carbon, which is then blown out past the exhaust valve."[66] A few owners devised homegrown or folk variants on the chain technique, like the man who reported success on dropping "a few bicycle ball-bearing balls, about 3/16 of an inch in diameter," into each cylinder, firing up the engine, and then waiting for the ricocheting spheres to knock the stubborn carbon deposits free and remove them with the exhaust.[67] Unlike brass or copper chain, hardened ball bearings could score and damage piston heads and cylinder walls, but the water technique was perhaps the least injurious to engines. It required owners to run up their engines to a high speed with the muffler cutout open (a device that allowed exhaust to bypass the muffler, then a common accessory but one that has long been illegal), and then to shove a water hose into the carburetor's air intake. The water blasted bits of carbon free from cylinder walls and pistons, advocates claimed, causing it to exit through the cutout without clogging the muffler.[68] Motorists also tried a host of "decarbonizing liquids, powders, and pills" pedaled by inventors, each of which claimed to be "quite effectual in loosening carbon deposits."[69] Some believed that kerosene—itself "a flagrant cause of carbon deposition," in experts' eyes—would dissolve the pesky substance. "About once a week," one believer advised, "if the car is used a good deal, put a couple of tablespoons of kerosene into each cylinder" and leave it overnight. On starting the car the next day, he promised, all the carbon would be gone.[70] Also competing for the motorist's dollar were mechanical devices that allegedly would provide a technological fix for the carbon problem. One was the "'Oxygenerator' Carbon Killer." It easily bolted onto a car's carburetor, and in 1916 its inventor claimed some 6,753 units were "now in use."[71] Yet none of these remedies eliminated the troubles caused by carbon; only with better understanding of combustion and improved engine and carburetor design was the pesky problem ameliorated.

Meanwhile, motorist and mechanic alike acknowledged that, as *Fordowner* put it in 1914, there is no more important part of the car than the carburetor. "One motorist calls the carburetor the heart of his car, another its lungs," the magazine noted, but the carburetor played a crucial role in vaporizing fuel for internal combustion engines. "There is no part of the car," *Fordowner* concluded, "and none more subject to vital ailments" than the carburetor. It could be responsible for excessive carbon buildup, difficult starting, poor acceleration, rough idling, lack of power and speed, poor gas mileage, and an engine's mysteriously and suddenly quitting.[72] One motorist's experience with this last ailment underlined the fussy quality of carburetors. His engine "suddenly developed a habit of shutting down without any apparent cause" because the pinhole-sized carburetor jet through which fuel entered became clogged with "goo" or "slime," a result of road dust mixing with gasoline. A single, tiny particle could wholly obstruct the narrow jet, causing the engine to cough, balk, and suddenly stop. Most puzzling, he observed, was that "the trouble could disappear as quickly as it occurred," for by the time the engine had quit and the operator investigated under the hood, "any heavier sediment or grains of sand might settle" in the carburetor bowl in which case the engine would start again after a short rest. It would run only briefly, however, before "the flow of the [gasoline] feed would pick it [the sediment] up and stuff it into the jet," once again stalling the car.[73]

Besides being delicate, precision devices, carburetors were designed to be adjustable. And given their importance to an engine's performance—to its starting, idling, acceleration, and fuel consumption—owners were inclined to think, whenever their car wasn't running well, that their carburetor needed adjustment. Carburetors were small devices, about the size of an orange or a couple of lemons, and always accessible, sitting either on top of the engine or mounted on one side; open the hood, get out a screwdriver, and the carburetor could readily be tinkered. "It may be all right," a Port Jervis, New York, man wrote to *Automobile* in 1911, "but I cannot help tinkering with the carburetor."[74] One expert dubbed this condition, "tinkeritis," an "expensive and nerve-racking disease" and urged that "at the discovery of the very first symptoms, the patient should consult with an experienced motorist, heed his advice, and thus stifle it in its early stages."[75] Another automotive authority explained that it "is very easy to spoil a good adjustment by random tinkering" on the carburetor, "and much less easy to restore it."[76] Automobile makers sought to prevent owners from tinkering. "A piece of advice," the 1909 Franklin manual told owners: "consider your motor as you do your watch and do not try to change it." Cadillac pleaded with the owner "who is constantly tinkering with his car when there is no necessity for it. The car left the factory in perfect tune and the carburetor required no further adjustment," the company explained. "We warn you against taking the carburetor apart and trying to fix imaginary troubles," stated a Locomobile manual, while one for the 1921 National "Sextet" resorted to boldface type as it screamed "DO NOT TINKER" to get readers' attention, calling tinkering "the cause of as much trouble as neglect and abuse."[77] Zenith, a major manufacturer of carburetors, inveighed most strongly against tinkeritis, likewise imploring readers of its manual in boldface: "We ask you to believe that every part of the carburetor is well designed for its purpose and should not be changed or modified except as indicated in our instructions." Never should the carburetor jets be altered, as "they are carefully gauged at the factory, tested for flow of gasoline and brought to a standard size according to the number, stamped on each." Finally, it cautioned owners to never make *any* adjustment unless they "knew beforehand the reasons" for doing so.[78]

In spite of all the warnings and admonitions, the carburetor long remained a kind of war zone, a contested site on the automobile where owners, frustrated by their car's poor performance, battled the manufacturer, who sought to defend the delicately calibrated device against being put out of order. The carburetor *was* responsible for much car trouble, so it was not unreasonable for motorists to think a small adjustment would cure their problem, although some naively expected a magical tweak might improve acceleration, starting, hill-climbing, and gas mileage simultaneously. Some manufacturers countered by redesigning carburetors to make them less adjustable, but that strategy proved problematic. Discovering such a carburetor on his new car, a Philadelphia motorist complained about the "non-adjustable nozzles." He acknowledged the "tendency on the part of inexperienced purchasers to get the adjustable type of nozzles out of order," yet pointed out the undesirable consequences of the new design.[79] "Often in making a machine 'fool-proof' (insulting phrase)," he observed, "the tendency is to produce a machine which, while it will run tolerably well for everybody, will never run just right in the hands of anybody." He captured precisely the dilemma the manufacturer faces in producing any complex personal technology for the mass market. To be user-friendly and widely adopted, a machine should neither demand nor permit too easy adjustment, especially if those adjustments depend upon a considerable command of engineering theory, as is the case with carburetors, and if returning the device to its previous settings might be difficult. Yet if the device severely limits the ability of owner-users to adjust it for local conditions, its performance will be unacceptable. The industry essentially agreed, and carburetors remained easily adjustable. As late as 1924, a writer claimed "tinkeritis has put more cars in the 'hospital,' caused more grief, worry and expense on the part of the car owner than all the natural wear and other troubles combined." Yet that was about to change.[80]

THERE IS CONSIDERABLE EVIDENCE that by the late 1920s owners were tinkering less with their cars, a trend that continued even amid the Depression of the 1930s. This decline of tinkering can be read in the changing rhetoric of advertisements. Back in the first two decades of the century, ads regularly touted the ease with which owners could get at their vehicle's tools and access its mechanical components. "You may look at the *outside* of your car before you buy it," a 1903 Haynes ad had noted, warning prospects about the allure of skin-deep beauty, "but you are sure to look at the *inside* before you have had it long."[81] Advertisements often bragged about the ease with which a motorist could get "inside" a car, that is, into its mechanisms. "Pronounced Accessibility is a Feature of this Scientifically Designed Power Plant," the ad for a 1910 Stevens-Duryea asserted.[82] By the 1920s, however, references to accessibility had disappeared from ads, suggesting that it had become either less necessary or of reduced concern for car owners. Similarly, owner's manuals no longer provided much information or advice regarding "roadside emergencies"; a 1926 Flint manual noted how "present day conditions are such that little difficulty is experienced even over long motor trips with automobiles," indicating motorists had little need to tinker.[83] Similarly, factory-supplied tool kits began shrinking in the late 1920s and early 1930s. A "recent check" of Studebaker customers in 1930 revealed "that the average owner had no occasion to use more than the jack and plyers [sic] from his kit," observed Studebaker Service, the company's in-house magazine for dealers and mechanics, so no longer would car buyers receive a tire pump, grease gun, socket wrench, hubcap wrench, and some of the open-end wrenches that once had been standard equipment. "The general availability of all kinds of automotive service," Studebaker Service's editor wrote, "and the universal practice of carrying a spare tire," by then really a spare tire mounted on an extra wheel, "renders much of the tool equipment furnished with cars of previous production unnecessary."[84]

By the end of the decade, "all kinds of automotive service" could be obtained virtually everywhere, a development that removed much of the necessity for car owners to tinker themselves. Whereas early motorists in the 1900s sometimes had to ship their cars hundreds of miles if they wanted factory-trained mechanics to work on their machines, thirty years later most small towns in America had at least one garage, auto dealership, and filling station.[85] Already by 1910 specialized gasoline filling stations had emerged, quickly supplanting fuel sales from hardware stores, animal feed handlers, and other general retail establishments. Ten years later there were some fifteen thousand gas stations in the United States, and about 8 percent of them not only sold fuel along with automotive parts and supplies but also employed mechanics to deal with customers' car troubles. As the use of automobiles exploded during the twenties, the number of gas stations increased nearly tenfold. Sprawling along every commercial strip and even infiltrating residential neighborhoods—where they often disguised themselves as colonial or Spanish colonial architecture—nearly all of the 124,000 gas stations existing as the nation slid into the Depression not only sold fuel but also maintained and repaired cars.[86]

New car dealers also were "selling service," that is, courting owners to bring their cars back to them instead of to independent garages and service stations for maintenance and repair. Dealers in the twenties increasingly advertised the benefits of "factory authorized" service and "genuine" or "authentic" spare parts, part of an effort to help their bottom line in a saturated and stagnant new car market. They also redecorated their facilities, covering dark and dingy garage spaces with white paint and sprucing up customer waiting rooms to attract service customers. Such cosmetic changes did little to improve the quality of service, suggests historian Steven McIntyre, yet the increased availability of professional maintenance and repair service in the period further undercut do-it-yourself tinkering.[87]

Improved automobiles, however, contributed the most to the decline of tinkering. By the mid-1930s, a new car was a vastly more robust and reliable machine than its predecessor of ten, twenty, let alone thirty years earlier had been. During most of those thirty years the pace of innovation had been explosive, and almost annually the industry's new offerings had relegated earlier vehicles to obsolescence. As early as 1906, an observer could assert that "the modern automobile does not give half as much trouble" as a machine "of yore." His perspective, while accurate, was also relative, yet over the years once frail and finicky horseless carriages evolved into more sturdy and dependable, and more user-friendly, transportation.[88]

Many of the innovations underpinning this transformation—electric starters, synchromesh transmissions, all-steel bodies, and alloy springs and axles—have already been mentioned. Just as important, however, were incremental improvements in the materials making up automobile components and in methods of manufacture. Advances in rubber chemistry and fabrication, for instance, went a long way toward eliminating the early automobile's easily damaged tires. Initially made of soft rubber that was readily cut or punctured by stones and gravel, dissolved by gasoline and oil, made brittle by sunlight, and weakened by water, modern tire chemistry eliminated those weaknesses. During the 1910s the part of the tire making contact with the road was no longer smooth but carried treads, affording greater protection against damage and puncture as well as better traction. By the end of the 1920s, as noted, balloon tires supplanted the narrower and larger-diameter ones of earlier years. With more tread contacting the road, balloon tires provided even better traction and, inflated to only thirty or thirty-five pounds per square inch, were not under as much strain as earlier tires. Flats and blowouts did not altogether disappear, of course. Prohibition proved especially injurious to tires as "dead soldiers," or broken whiskey bottles tossed from car windows, littered highways. The mileage motorists might expect to get from their cars' tires, about fifteen hundred miles at best around the turn of the century, steadily increased, with tire makers guaranteeing thirty-five hundred to six thousand miles by 1919 and up to eight thousand by 1922.[89] When a modern car's tire did go flat or blow out, changing it was much easier, nothing like the ordeals motorists endured in the first two decades of the century. By the thirties, as noted, the practice of carrying a spare wheel with tire already mounted and inflated had become common.

While most motorists were aware of the improvements in tires, few realized the hidden, even invisible, automotive applications of rubber that also increased automobile longevity and reduced the need for tinkering. As chemists learned to make rubber less susceptible to damage from petroleum products, rubber belts supplanted leather ones to transmit power from their engine to cooling fans, water pumps, generators, and other mechanisms under the hood. Rubber belts stretched less than leather and did not slip when wet; they also ran more quietly and lasted longer—"it is not unusual now for owners to drive their cars a year before it is necessary to replace a fan belt," noted a Ford dealer in 1924.[90] Similarly, fuel-resistant rubber coverings protected ignition wires and spark plugs from water, reducing the chance of short circuits and ignition trouble during rainy or snowy weather. Rubber seals and gaskets also kept water out of exterior light fixtures, minimizing short circuits and rust, and sealed the connection between carburetor and air cleaner, all but eliminating the buildup of goo and slime that had clogged earlier carburetors. Underneath the car, rubber "boots" on shock absorbers and other components kept out dust and grime.

Innovations in automobile assembly were no less significant. Whereas early motorists had feared "losing" their automobile piece by piece as bolts shook loose, the development of standard-sized lock nuts and washers meant that components bolted on generally stayed put. Manufacturers also shifted away from the use of bolts or rivets to the use of welding for fabrication of the chassis, and by the mid-1930s most automakers, led by Dodge, also produced bodies entirely of welded steel, a construction that was infinitely stronger, more crash resistant, less noisy as it aged, and easier to repair than that of older bodies in which metal panels covered wooden frames.[91]

Finally, the automobile carburetor, that once major cause of "tinker-itis," also benefited from technical improvements. After a long cross-country trip with her husband in 1924, motorist Mary Bedell reported that "we had no need of adjusting our carburetor at any time." Even cross-ing the mountains, she added as if surprised, the car "gave us no trouble at all by getting short-winded."[92] The authors of a 1927 textbook on auto mechanics felt obliged to remind their readers "that carburetor adjust-ments are not the cure for all troubles found in the automobile engine."[93] Indeed, as a historian of automotive technology concludes, "Probably no part of a modern car has undergone more thorough development in the period around the 1920s" than the carburetor.[94] During the same years, automakers introduced what were called cam-actuated fuel pumps, the first wholly reliable means for getting gasoline from the tank, wherever it was located, to the carburetor. This innovation rendered the older methods whereby fuel was delivered to the carburetor—hand-operated air pumps or simple gravity feeds, a problem on steep hills—obsolete and made possible the replacement of old-fashioned, updraft carburetors with more efficient downdraft models. Coupled with the incorporation of automatic carburetor chokes that made starting a cold engine easier, all of these improvements in fuel systems reduced the need for tinkering and adjustment that was required with older vehicles.

As the automobile became more reliable, it ceased to play a role as a demanding mechanical tutor. By the latter 1930s, motorists had little need or incentive to study their machines as had Philadelphia's Miss McGaw in the first decade of the century who deliberately tinkered under the hood to prepare herself for roadside emergencies. Nor were lengthy road trips enforced tutorials in car trouble, like Winifred Dixon's in the early 1920s. "What I knew about the bowels of a car!" she exclaimed when it was over, but like most others she would have gladly forgone the greasy and labori-ous lessons.[95] When motorists finally had that choice, most ignored the bowels of their machines and as a consequence knew little about what was under the hood. Tinkering and tool use, once inseparably linked to owning and operating an automobile, had become a matter of choice. Those who still tinkered tended to do so in their leisure as a hobby.

It is not just coincidental that the decade in which new cars became "modern," the 1930s, was the same one that saw the start of the hobby of restoring and collecting antique vehicles. In 1935, a handful of men, most old enough to recall the earliest days of horseless carriages and aware those ancient machines were fast disappearing, came together to found the Antique Automobile Club of America (AACA). It and its members comprised the first of many organizations that today make up the old-car hobby. Those hobbyists, then and now virtually all men, searched old barns and junkyards for restorable cars and, like collectors generally, bragged about their finds: the low-mileage car in running condition that had sat in a barn for twenty or thirty years and was perfectly preserved; the rare model restored from a nondescript pile of rusted parts salvaged from a farmer's field or old garage; the proverbial basket case in which instance the hobbyist might literally have gathered the parts up in bushel baskets. After lovingly restoring their old car to shiny newness, often spending years in the process, these hobbyists relished recovering the forgotten skills of cranking engines, double-clutching to shift gears, lighting pilot lights on steamer boilers, or fiddling with acetylene generators to throw light on the road ahead when driving at night. Having reversed automotive progress and returned, so to speak, to the technological past, these old-car enthusiasts practiced the sort of tinkering modern motorists had pushed toward extinction.

Antique auto enthusiasts also enjoyed operating their "new" old cars, entering them in shows and competing for awards in categories by age, type, or marque along with the invariable prize for "best in show." Behind the wheel (or tiller) of their antique vehicles, they proudly carried dignitaries or attractive women in local parades, often attired in period costumes. Or they toured in company with other club members through the countryside, turning heads as a stream of twenty- or thirty-year-old traffic passed through a community. Some hobbyists reenacted historic long-distance trips made by motoring pioneers, such as AACA member George C. Green, who in 1938 drove his restored 1904 Curved Dash Oldsmobile from Philadelphia to California and back. "The Oldsmobile gave almost no trouble," Green reported, and on his best day he and his wife covered 240 miles, a reminder of how much improved highways contributed to user-friendly motoring.[96]

Invisible to 1930s spectators admiringly ogling lovingly restored vehicles as they passed in a parade or through town were the technical texts—the service bulletins, factory-issued shop manuals, and most notably the owner's manuals—that old-car hobbyists had sought out to enable them to resurrect, restore, and run their antique machines. Whereas modern motorists tended to leave their owner's manual where it was when their car came from the dealer, unopened and unread in the glove compartment, anybody restoring an old car usually begins by procuring and studying the original manual (or a facsimile) plus whatever other contemporary technical literature can be found. Even for those who knew a lot about modern cars, "antique" manuals and technical texts were essential sources of information about how those vehicles were constructed, about their particular parts and components, their performance, and for tips on starting and operating them. In the 1930s, to climb into the driver's seat of a 1904 Oldsmobile, 1911 Maxwell, or 1924 Model T Ford was one kind of time travel; another occurred as old-car hobbyists opened the pages of the original owner's manual or any other technical publication to reenact the same learning process that those who were the first-time buyers of such cars had performed years earlier. It is to such literature, and to the learning it contributed to the adoption of a new and inordinately complex technology, that we move in the following chapter.

Chapter 5

Reading the Owner's Manual

I T IS OUT OF THE QUESTION TO SEND AN EXPERT with every vehicle shipped," explained the editor of *Motor Age* in 1902, "but the instruction book can go as a silent instructor."[1] His comment captured an essential truth: unlike purchasers of traditional goods, technology consumers desperately needed help—in learning to use a complex machine, in diagnosing its troubles, and in getting it working again. The solution was the silent instructor, or owner's manual. Representing a machine in words, diagrams, and illustrations, an owner's manual created a simulacra of the technology itself. Readers could study the printed (or more recently, on-the-screen) version of their technology whenever they had questions or problems.

The practice of including instructions with machines had started in the early nineteenth century, and, as noted in chapter 1, by the time of the Civil War, with the advent of sewing machines, the brief labels that came with clocks had morphed into small booklets, the first true owner's manuals. By the time the horseless carriage appeared at the turn of the twentieth century, consumers had come to expect instruction books with any machine they bought. If an automaker inadvertently forgot to supply a manual with every car, as Studebaker once did in the 1920s, purchasers raised a ruckus.[2]

Instruction books were essential because of the automobile's complexity. Cars had many more individual parts than the few dozen or hundred of earlier consumer technologies, and their components were subjected to greater stresses in ordinary usage and, as we have seen, often broke down. Finally, because some of the car's systems were electrical and chemical rather than just mechanical, as on earlier personal technologies, consumers struggled learning the new knowledge required to operate battery-powered ignitions or acetylene-gas-illuminated running lights. The relationship between automotive technology, new learning, and instructional texts was summed up cleverly by Andrew L. Dyke, an early advocate of horseless carriages, erstwhile automobile manufacturer, and publisher. He reformulated the old cliché linking knowledge and power to read, "Knowledge is Power (H.P.)," the letters standing for "horse power." Coined to promote a handbook he compiled, *Dyke's Diseases of a Gasoline Automobile and How to Cure Them*, the aphorism meant that only by acquiring new knowledge, often by reading, could one obtain "power" from or over an automobile or any other complex personal technology.[3]

Consumers often found the first manuals automakers provided to be inadequate and demanded additional information. "If you have not already done so," Charles Henry wrote the maker of his new 1905 Winton automobile, "please send me some kind of directions how to adjust the new ignition battery and all about it; also directions in regard to the oiler and the carburetor."[4] Another early automobile owner claimed his instruction book was so opaque that it might as well have "been printed in Choctaw or Sanskrit, so far as understanding its quiz-compendious manner" was concerned, a reference to the way early manuals were sometimes organized like catechisms, providing answers to randomly posed questions.[5] Even the editor of *Motor*, an early automotive periodical, believed that "better instruction books" were imperative if "a man who does not know the exact difference between a four-cycle engine and an eight-day clock" was to learn how to operate a motor vehicle.[6]

Confused and frustrated, motorists often sought out more "definite and understandable information" in works authored by people unconnected to the automobile manufacturers.[7] One of the earliest such volumes was Dyke's *Diseases of a Gasoline Automobile*, which first appeared in 1903 and remained in print until the 1940s, going through many editions. During the first quarter century of the auto age, at least seventy-five such independently written volumes were published in the United States. The popularity of books like *The A-B-C of Automobile Driving* (1916); *Auto Repairing Simplified* (1918); *First Aid to the Car* (1921); and *Everyman's Guide to Motor Efficiency* (1920) signaled the consumer's craving for more accessible, easily understood information.[8] One of these volumes,

Gasoline Automobiles (1915), was a "technical best seller" for publisher McGraw-Hill, selling 83,500 copies in its first eight years.[9] Another major publisher of technical literature, Norman W. Henley, of New York City, had an even bigger hit with *The Model T Ford Car*, also published in 1915 and going on to sell nearly a million copies during its first eight years, or roughly one book for every ten Model T owners.[10]

Besides owner's manuals and independently published handbooks, motorists gained valuable knowledge from magazine articles. Early adopters read *Horseless Age*, the trade journal that first appeared in 1895 and which targeted auto owners and potential buyers. By 1910, the appeal of motoring among the general public had prompted general circulation magazines as well to run how-to-do-it articles for motorists. One of the first was *Country Life*, a magazine that catered to affluent suburbanites, a group in the vanguard of adopting horseless carriages.[11] Articles like "Learning to Drive a Motor Car" (1907) and "Common Sense in Automobile Driving" (1908) started a trend that rapidly spread to other general interest periodicals. In 1912 *Literary Digest* published "The Mistakes of Beginners"; in 1917 *Ladies' Home Journal* ran the piece "Little Things about a Car That Every Woman Who Means to Drive One Ought to Know"; while in 1924 even the venerable *Atlantic Monthly* printed an article titled "Things That Every Owner and Operator of Motor Vehicles Should Know."[12]

People turned to books and magazines not only to better understand their automobile and its operation but because, as historians have argued, reading had taken on new cultural salience by the early twentieth century. The cultural historian Warren Sussman wrote about how Americans, as they moved from farm to towns and cities, consulted how-to-do-it books on etiquette, cooking, and home decoration to familiarize themselves with urban ways and better adapt to those conditions. Similarly, Susan Strasser has shown how "cooking by the book," that is, following recipes stipulating precise quantities and exact cooking times, also increased during those years. And generalizing more broadly about the newfound authority Americans accorded the printed word, Burton J. Bledstein has suggested that "buying a guidebook, manual, handbook, or book of reference in order to function—*by the book*—became a necessity of everyday activity for the middling classes."[13] Motorists felt the same necessity, turning to the printed page—be it in their owner's manuals, motoring

handbooks, or articles in automotive or popular periodicals—to understand their automobiles "by the book." Learning by reading was in fact
inseparable from adopting the new technology.

Manuals, handbooks, and other prescriptive technical advice directed
to early automobile users open a window on the problems they confronted, and in previous chapters I have often quoted such literature for
that precise purpose. Here I look at these technical writings differently,
not so much as sources but as historical actors in their own right. My aim
is to examine "the role of the word in making the world," as historian of
science Charles Bazerman felicitously phrased it.[14] Bazerman was writing
about the importance of the word to the emergence of scientific practices
and method in the seventeenth and eighteenth centuries. The word continued to play a critical role, often accompanied by images or diagrams,
in the world of technology consumption in the nineteenth and twentieth
centuries. Technical literature helped consumers choose which machines
to buy and taught them to operate and maintain them. In examining the
technical writing directed at car owners, we will see how its content, aims,
and rhetoric changed over time in response to the evolution of automotive technology, the environment in which vehicles were used, and the
demographics of automobile ownership. We will look at two genres: the
owner's manuals that came with new cars and mass-circulation magazines that published articles imparting advice about motoring.

From around 1900 to 1925, manuals addressed what may be called the
"operator-mechanic," an individual who was, as we've already seen, likely
to be but not necessarily a man and definitely a novice when it came to
automobiles. Because these consumers were making the leap from the
horse-and-buggy era to the auto age, manuals instructed them about their
car's workings, operation, and the necessary maintenance and adjustments that would have to be performed to keep it running. As automobiles became more robust and reliable and people grew up living with
them and thus familiar with them, owner's manuals shifted their emphasis. By the late 1930s, they spoke to what we might term the "driver-pos-
sessor," a man or woman who most likely had already owned an earlier
model car, knew how to drive, and no longer needed to be tutored in its
mechanical workings. These manuals instilled pride in car ownership
along with tips on use. They advised owners to take their vehicles to the
dealer's service facility for all maintenance and problems. Thus, manuals

no longer provide detailed information about a car's mechanical systems or presumed owners would work on or maintain their cars themselves.

During the first generation of motoring, however, an owner's manual had been as important to operator-mechanics as were the spark plugs, radiators, steering wheels, batteries, or transmissions of their vehicles. Confronting an incredibly complicated machine, consumers needed the knowledge conveyed by the manual, a fact manufacturers impressed upon them. "The information, advice and instruction contained in this book are published because the user of a motor car needs them," explained one Cadillac manual, while another asserted that "this book is a part of your car, and should always be kept where it is readily accessible."[15] The cover of an early Oldsmobile instruction book admonished the reader: "Keep this book in your tool box. A duplicate will be furnished only upon satisfactory reasons being given for replacing same."[16] Because automobiles incorporated components or subsystems manufactured by many different firms, buyers often received more than one manual. In the "tool equipment box" of the 1920 Franklin could be found "instruction books for the Atwater Kent ignition system, the Stewart vacuum system, the Willard storage battery and an instruction card for the 'Lift-the-Dot' fasteners [used to attach and detach the removable side curtains and folding top]."[17]

In spite of the operator-mechanic's need to read and learn from manuals, it took the anonymous authors of those texts nearly twenty years to figure out how to write comprehensibly and effectively for novices. Occasionally, authors seemed aware of the tremendous challenges facing first-time automobile buyers, but more often they produced manuals that only an automotive engineer or other experienced technician might parse. Early manuals were often of puzzling brevity, without tables of contents or indices. They referred to car parts using technical terms that were neither defined nor illustrated, making it all but impossible for inexperienced readers to figure out what was being discussed.

Consider *A Study of the Oldsmobile, A Little Booklet for Users of the Oldsmobile Curved Dash Runabout,* the most widely distributed silent instructor of the pioneering era. Accompanying the 1901–4 Oldsmobile, a two-seat, tiller-steered vehicle that cost only $650 and that sold over ten thousand units, justifying the historian's label "first quantity-produced car."[18] The manual's thirty-two pages make clear the challenges novices—

those who did not know the "difference between a four-cycle engine and an eight-day clock"—faced in trying to learn from such texts.

The manual starts on an appropriate introductory note, assuming the reader knows nothing about a car or how it works. "Let us step to the back end of the machine," the text says, "and placing our two hands under each side of the back, turn the knobs; this will relieve the top, or 'deck' of the body, which we will lay to one side out of the way. Also drop the tail gate by turning the 'T' handle at the back of the body. Stepping up to the front of the machine we will remove the cushion and lift up the bottom of the seat, which practically exposes the whole of the inside of the machine (See Fig. 1)."[19]

The tour of the machine continues as the text directs the reader's attention to components now visible with the body removed: the chassis frame, motor, gasoline tank, water cooling system, petcocks to drain the cooling system, and so forth. Brief asides offer lessons about the construction, purpose, or operation of different parts: "At the right we see the gasoline tank, which holds five gallons" or "Attached to the upper pipe we see a small pump at the end of the shaft to lift and circulate the water [in the engine for cooling]." Some of these lessons are warnings: "The long clutch lever or controller lever (6CB7-1008) connects the motor with the vehicle and should never be moved when the motor is running, unless it is desired to start or stop the vehicle" (7). This statement describes the basic function of a clutch, but because it appears early in the manual before the discussion of how to drive the car, may have fallen on deaf ears. Moreover, the sentence only hints at the fact that touching the lever at the wrong time could damage the car's delicate mechanisms. Most confusingly, the manual introduces novices to their new Oldsmobiles using sentences bristling with technical jargon. "When the case-hardened cone of the high speed clutch is thrown in," as the introductory section concludes, "no gears are running, and the only bearings in use are the three heavy main bearings of the crank shaft itself" (8). Never does the manual define terms like *case hardened cone, main bearings,* or *crank shaft.* Understanding "the whole of the inside of the machine" with this manual as ones only guide was difficult indeed.

Although it has become standard for technical manuals to rely on illustrations and diagrams to explain machines and instruct people in their use, *A Study of the Oldsmobile* was logocentric, including a mere three illustrations. The first was a line drawing of the car's chassis as seen from above, with the body removed. In a nod toward making the drawing intelligible to laypersons, the illustrator had provided a hint of perspective by shading the tires to suggest they were round, but overall the illustration resisted easy understanding. It was untitled and had no caption or label, save for two unexplained words—*relief button*—in the middle of the chassis. The drawing probably is what the text quoted above was referring to when it said to see figure 1, but without any caption, labels, or title, readers who were unaccustomed to technical manuals would have had little way of knowing for sure. And of course without labels there was no way for them to usefully relate what they looked at on the page to what they saw on their car. The other two illustrations the manual provided were even worse, as they were mechanical drawings of the kind engineers use, lacking perspective and even harder for a layperson to understand. One depicted what appears to be a cross section of the car's clutch and gear mechanism and had numbers superimposed on the drawing that corresponded to a list of parts on the opposing page. The second illustration showed a side view, or elevation, of the Oldsmobile's one-cylinder motor, but it too was untitled and lacked labels. Nowhere did the text refer to the list of parts and the drawing it related to, or to the drawing of the engine. In short, whoever oversaw the manual's publication knew illustrations would be helpful for conveying technical knowledge but not how to create ones that would communicate useful information.

Halfway through the manual, the reader found the first instructions on how to prepare the Oldsmobile for use and how to operate it. "See that there is plenty of gasoline in the gasoline tank and water in the water tank" (8), a new section begins, assuming former horse-and-buggy users knew nothing about feeding an automobile. Similarly, the text stressed the importance of lubrication: "If you allow one bearing on your machine to run dry, or cut," the manual warned, "you will do the machine more harm than a year's use would" (9). Readers were even told on which side of the two-person seat to sit—not an obvious matter, given that the steering tiller was centrally mounted and could be manipulated as easily from either seat. "Take the right-hand side of the seat and see that the controller lever (6CB7-1008) is thrown clear back next to the reverse gear," the text instructed before ticking off tasks preparatory to starting the motor (12). Starting required "turning the starting crank (6CJ1-1015) over two or three times quickly" (13). Given the car's construction, only if operators were positioned on the right side of the seat could they reach this crank, as it protruded through the right side of the Oldsmobile's body just below the seat.

Because starting early gasoline engines could be devilishly difficult, the manual attempted to address the situation "when [the] motor does not ignite regularly" by saying, "it may be due to one of five things."

First—Not enough, or too much, gasoline.

Second—Loose wire connections on battery or engine.

Third—Short circuit at ignition plug.

Fourth—Changing the adjustment of vibrator on spark coil.

Fifth—Weak battery; this is not likely unless much used for four or five months. (30)

This analytical, checklist approach to diagnosing problems followed practice that had long been employed by the medical profession, and within a decade this kind of thinking—if not *a*, then *b*, and if not *b*, then *c*, and so forth—would become the primary diagnostic method both for auto mechanics and for operators seeking the cause of car trouble. The only problem with such checklists as the one quoted here is that their items were not adequately explained for inexperienced readers, leaving beginners to discover for themselves how much gasoline was not enough or too much. Similarly, they would need to figure out how to locate and test for a "short circuit at ignition plug" or how to set and adjust the vibrator on the spark coil (assuming they even knew what the device was in the first place and where to find it).

The Oldsmobile manual's section titled "Operating the Oldsmobile" included instructions for shifting gears, stopping the car, and managing the machine while under way. These were the very first instructions to the novice included in the manual about what we would call driving, but they were not always easy to understand. Climbing hills was one challenge. "If you have the speeder [throttle] at full speed, and the road is good so that the motor would run still faster," operators were instructed, "push the spark lever ahead" in order to avoid "pounding." But the text never told readers how to recognize "pounding," nor, for that matter, what the "spark lever" was and where it could be found on the car. When moving "upgrade, so that the motor pulls to a slow speed," the text continued, "change the sparking lever backward again to an incline position; too early a spark with slow speed will cause the engine to work against itself and stop." Heading down hills, readers were advised to, "when nearing the bottom of the hill apply the speeder to give the motor as great speed as possible in order to make the next hill" (14), a useful tip for operating a low-powered, one-cylinder car.

Considering the operation of the Oldsmobile, readers might intuit from the nearly three dozen sentences in the manual prefaced by the words *never, always,* or *don't* that they might easily damage the vehicle by improper or inattentive operation. Warnings such as "Never use very late spark continuously when on the road" or "Never allow [the] motor to run slow when greater power is needed" underscored the vulnerability of the machine and implied the operator should be attentive to the vehicle's many needs. Of all the manual's admonitions, only two suggested that carelessness might result in human death or injury. "*Never* fill [the] gasoline reservoir by *lamp light,*" commanded one, while the other read, "Never make a quick turn of the steering lever while the vehicle is running at a high speed; it is liable to cause a bad accident" (31–32). Because, as noted in chapter 4, the word *accident* in the first decades of the century commonly referred to mechanical breakdowns rather than incidents involving physical injury to vehicles and their inhabitants, readers might miss the more important point here, that a quick turn could flip the Oldsmobile and break somebody's neck; only the adjective *bad* provided a clue as to this more pernicious meaning. Surely authors of owner's manuals knew automobiles could be extremely dangerous, but the rhetoric of "don't," "never," and "always" was almost always deployed to protect machines, not human beings. Manufacturers wanted to sell cars, and their literature rarely addressed the dangers of speeding vehicles or negligent driving.

But every owner's manual well into the 1930s did address the subject of car trouble. The manufacturer of the Oldsmobile assured purchasers that "we only make one kind—the motor that motes"—a play on the verb *locomote*—but devoted fully a third of the pages in *A Study of the Oldsmobile* to situations where the car failed to mote. But except for the situation where the motor "fails to ignite," discussed above, the manual provided only the sketchiest hints for diagnosing difficulties, and then little advice as to how to remedy them. Its treatment of trouble with the carburetor, or "mixer," as it was often called, was typical. "If it does not work properly," the text explained, "it is because 'something has happened to it,' or the motor valves are not properly set" (13). Novice operator-mechanics needed more than this, perhaps a list of what to look for in checking the condition of their mixer or some discussion of the things that might have happened to it. The manual's shortcomings must have exacerbated the frustrations of being an early adopter of the new transport technology.

WHY WERE EARLY AUTOMOBILE OWNER'S MANUALS so lacking in detail, adequate explanations, and helpful illustrations? In comparison, the printed instructors sold a generation or two earlier with sewing machines were generally far more comprehensive, clearly organized, pedagogically inventive, and better illustrated. The explanation of the differences between the manuals for sewing machines and those for early automobile is gender: automobile manuals were written for men while those for sewing machines were addressed to women. According to the student of technical writing John Brockmann, it was the female market for home sewing machines that prompted the industry to produce manuals in the first place.[20] Their anonymous authors wrote for a middle-class female whom they presumed was not only ignorant of machinery but whose nature was antithetical to the mechanical arts. Owner's manuals responsive to the needs of such women became important in persuading them to purchase and bring home the complicated new machines. It is intriguing to speculate whether women influenced the writing of sewing machine instructors, either directly as editors and authors or indirectly through discussions with their brothers, fathers, or husbands who might have been involved with such texts. Certainly women, as mothers and the mentors of children, had more expertise than did men in teaching beginners and anticipating their difficulties in learning new tasks.

The authors of automobile owner's manuals like *A Study of the Oldsmobile* were no less anonymous than those who wrote sewing machine instructors, but there is no reason to think any of them were women or even that women influenced these texts.[21] Early automobile manuals were clearly authored by men who spoke "Motor" and who shared the widespread stereotype that *all men* were equally interested in and knowl-edgeable about machinery, the mirror image of the stereotype holding that all women lacked mechanical aptitude. Automakers in those years had little incentive to improve their manuals, however, as they were able to sell virtually any cars they built. In the 1910s, however, as competition increased and sales mushroomed, the auto industry became more aware of the almost universal ignorance buyers had about the new technology and began to improve its manuals.[22]

Texts became more conveniently organized, usually with an index or table of contents, allowing readers to easily find things. In addition, in the 1910s manuals usually contained well-captioned visual aids and were written in nonspecialized language.[23] By 1915, most manual texts ran forty or fifty pages in length, though some were considerably longer, like those for the 1921 Oldsmobile and Oakland, another General Motors car, which contained 121 and 162 pages respectively.[24] Manuals addressed readers as true novices who needed tutoring as operator-mechanics. Clearly the authors and publishers of these manuals were themselves no longer nov-ices; they had learned to effectively communicate with neophytes who were moving from the horse-and-buggy age into the auto age. They had a financial incentive to do so; a good manual might help sell their brand of automobile over another.

The *Ford Model T Instruction Book,* the manual accompanying what by the early 1910s was the best-selling American automobile, typified the new approach. Only forty-six pages in length and roughly three by six inches, it could fit the pocket of a shirt or overalls more readily than the larger, roughly five- by seven-inch Oldsmobile manual. By dint of its accessible language and lucid writing, along with its well-crafted illustra-tions and diagrams, the Ford manual effectively conveyed a great deal of information in a space-efficient format. It assumed the Ford dealer would provide buyers with a basic introduction as to where controls were located—spark advance, starting crank, shift and brake pedals, and so on—but provided clear instruction on operating and maintaining the car. Two-thirds of the manual's pages addressed car trouble, defining terms that might be unclear to novices, such as *overheating* and *irregular ignition,* the latter being "the occasional miss in one or more cylinders." It also provided detailed, step-by-step advice to operator-mechanics who had to deal with problems on the road, templates for the sort of checklist thinking that enabled them to diagnose car troubles and remedy them. If the engine misfired, the manual explained, it probably resulted from ignition trouble that "can usually be traced to its source." To "ascertain which, if any of the four plugs, are fouled with oil, short circuited with carbon or inoperative from some other cause," the manual continued, readers should first

> open the [hand] throttle two or three notches to speed up the motor; now hold your two fingers on two outside vibrators [devices that rapidly opened and closed the ignition circuits] so that they cannot buzz. The evenness of the exhaust will show that the other two are working correctly and that the trouble is not there; or an uneven exhaust will indicate that it is between the two that are free. If the two cylinders fire evenly change the fingers to the two inside vibrators and again listen to the exhaust. Having ascertained in which pair the trouble is, hold down three fingers at a time until you find the one on which the motor does not fire.

By a process of elimination, then, readers would run tests to figure out which spark plug was problematic. "Remove the spark plug and clean the core," they then were told, and "replace plug, taking precaution that all connections are correct and right. If missing continues, put in a new plug." Because their trouble might not be due to a bad spark plug, however, the manual then told them to test their ignition coil, and if it were working, what to do next.[25]

The Ford manual also typified the way the auto industry had come to understand the old truism that sometimes a picture is worth a thousand words. It contained useful diagrams showing the car's wiring pattern; the "proper method for setting [the] commutator," a vital ignition system component; and a "fan belt diagram," illustrating "the proper method of adjusting the belt." Hands-on illustrations depicted over-the-shoulder views of operator-mechanics adjusting the bands of the planetary transmission and removing the main bearings and rods without having to tear down the entire engine.[26] Unlike the earlier Oldsmobile manual that described in words the essential task of lubricating the car, the *Model T Instruction Book* graphically showed readers where to oil or grease their vehicles. Like virtually all manuals from the mid-1910s through the

1920s, it included what came to be called a "lubrication chart." Captioned "Where to Oil the Model T," the Ford chart included a photolithograph of the car's chassis as seen from above. Surrounding the image of the chassis itself were blocks of text indicating the frequency of lubrication along with the type (oil or grease) for each lubrication point on the chassis. One text block read, "Rear Spring Hanger. Oil every 200 miles"; another, "Hub. Grease every 500 miles." To help novices find those locations, each text block was connected by a thin line running to the location on the chassis illustration that required lubrication.[27] Because the industry assumed most car owners would do this work themselves outside or in their home garage, lubrication charts were often larger than page-sized, like the one that folded out of the 1917 Oldsmobile manual. It was perforated so owners could tear it out, unfold it, and tack it up in their garage, as the text encouraged.[28] The large lubrication chart that similarly folded out of the 1926 Studebaker manual was printed in two colors, with red lines running from each point requiring lubrication on the chassis, depicted in a black-and-white photolithograph, out to a smaller close-up photograph of that same spot, thereby showing operator-mechanics the exact thing to be oiled or greased. Arrayed around the chassis on the edges of the chart, each little photograph was framed by a red line. These little framed images came in three different shapes, indicative of the frequency of lubrication: a square one meant weekly lubrication, a circular image monthly, and a diamond one every ninety days. The thickness of the red line framing the small photo indicated the type of lubricant: thin meant oil, thick stood for grease. Finally, the close-up photographs of lubrication points were all numbered and referenced in a key as on a map, identifying each by name. With a user-friendly visual aid like this, anybody could find their "Gear Shift Lever (Oil Hole)," say, and know when to lubricate it.[29]

In short, from about 1915 into the late 1920s automobile owner's manuals effectively tutored novice operator-mechanics in the ways of running and maintaining a car as well as diagnosing its troubles using step-by-step, deductive reasoning.[30] As the 1920s faded into the 1930s, however, owner's manuals once again changed, registering the fact that automobiles had become more reliable and easier to run and manage. By the end of the Depression decade, the transformation was complete, and as noted manuals now no longer spoke to operator-mechanics but to the driver-possessors who celebrated the pride and joys of automobile ownership but had little reason or necessity to pick up tools and tinker beneath the hood. This shift was evident in changes large and small.

The new generation of manuals gave far less attention to owner maintenance. The 1927 Lincoln owner's manual, for instance, eliminated the section that had told previous owners how to grind the valves of their engine. In the same year, Studebaker stopped handing out two- by three-foot lubrication charts for owners to tack onto the walls of their garages "due to the fact," as the company explained, "that a large percentage of the car owners realize that their car can be lubricated much more efficiently in the dealer's service station than in their own back yards."[31] Even *The Ford Model "A" Instruction Book,* the manual that accompanied the new model Ford that in 1928 supplanted the Model T, clearly conveyed the message that owners were no longer expected to do their own maintenance. The Model A manual not only contained fewer pages than Model T manuals but on its opening page advised owners that "when repairs or replacements are necessary, it is important that you get genuine Ford parts. This is assured when you take your car to an authorized Ford Dealer." The *Instruction Book* for the 1928 Hudson "Super-Six," similarly discouraged owner tinkering by explicitly distinguishing the relatively few "operations an owner should attempt" from those that "should be left to authorized Hudson-Essex Service Stations."[32] While manufacturers had economic reasons to discourage owner maintenance in the saturated car market of the late 1920s and 1930s, owners seemed happy not to have to get out the tools that came with their cars—the number of which was being cut back by automakers anyway.

Another change evident in the manuals is in their titling. Although I have referred to these publications from the outset as "owner's manuals," before the 1930s the word *owner* rarely appeared in their titles. Instead, they were called "book of instructions,"[33] "instruction book,"[34] "book of information,"[35] "reference book,"[36] or most commonly, just the "instructions."[37] Among the hundreds of manuals I examined, only one before the 1930s used the word *owner* in its title, the 1924 Nash publication, *Information for the Owner.* But as the celebration of possession and ownership supplanted content in manuals on maintenance and mechanical work in the 1930s, titles such as *1938 Buick Owner's Manual* and *Hudson 112, 1938 Owner's Manual* become common.[38]

Emblematic of the new emphasis in owner's manuals on pride of ownership is *Your New 1942 Nash,* which came with the few vehicles sold by the automaker in the fall of 1941 before Pearl Harbor closed down civilian automobile production in the United States. "You're the proud, practical owner of a 'Million Dollar Beauty,' a big, shiny 1942 Nash and you're starting out now on a wonderful trip," the first page announced, "a trip that may take you the equivalent of around the world—two, three or four times." On the same page was a large illustration showing a husband carrying a suitcase accompanied by his wife and daughter, about to get into their car. Beneath the image were the words: "Nash bids you Bon Voyage and wishes you Happy Days at home on wheels."[39] In the text's hyperbolic language, "wonderful trip" was to be taken literally as well as metaphorically, alluding to the long and satisfying relationship buyers would have as drivers and owners of a Nash.

That relationship began not with an introduction to the car's mechanical systems as in earlier manuals but a casual invitation for owners to get acquainted with their new companion. Nash did not even ask buyers to *read* the manual; "a few minutes with this book and you'll know what's going on under the hood," it cheerily suggested (4, 10–11). The Nash text also paid scant attention to driving skills, for now people picked up such knowledge growing up, learning from friends or family members. Nor did the manual contain any parts list, lubrication chart, or technical diagrams, and it hardly mentioned car trouble. Instead, the manual focused on appearance: "You're Proud NOW and you can STAY Proud with these 'House-Cleaning' Hints," it exclaimed in a two-page, detailed discussion of how to keep the Nash's interior clean and its exterior "bright and shiny." These included tips on removing upholstery stains caused by candy, fruit, blood, lipstick, and vomit (8–9). None of these substances was new in the 1940s, of course, and the last was probably a greater problem during the era of speakeasies and bathtub gin in the roaring twenties. Yet such stains took on new importance as the car no longer was a problematic and challenging machine but a "luxurious club lounge on wheels," a taken-for-granted but prized commodity. Appropriately, the longest single section in the 1942 Nash manual dealt with further shopping in that it was a catalog of "Nash Engineered Accessories." People could express their pride of possession by purchasing dashboard clocks, license plate frames, locking gas tank caps, seat covers, spotlights, and "the famous Nash Bed," which enabled the rear seats to fold wholly flat and link up with a "second mattress unit" in the trunk, creating a "home on wheels" (34–39).

Clearly, the owner's manual had come a long way from when it was to be considered "part of your car," as the Cadillac instructor had claimed itself to be some twenty-five or more years earlier. In fact, texts like *Your New 1942 Nash* had lost much of their reason for existence. They might still serve as a reference, although that role would only flourish in the future, in the postwar decades when automobile climate control systems, security and alarm systems, and entertainment and information handling systems became newly complicated and user unfriendly, forcing owners to read in order to understand their machines. By 1940, however, manuals no longer were vital to the ordinary running of the vehicle; they tended to remain unopened and unread in the glove compartment, where they came from the factory, marginal but not quite obsolete.

THE EVOLUTION OF PERIODICAL ARTICLES addressed to motorists paralleled the development we have traced in owner's manuals. Beginning around 1910, as the automobile became a popular topic of discussion, publications that previously had ignored motoring began running articles on the subject. Family and women's magazines like *Collier's*, the *Saturday Evening Post*, the *Ladies' Home Journal*, and *Good Housekeeping*, along with male hobby periodicals such as *Popular Science* and *Popular Mechanics* and even staid literary periodicals like the *Atlantic Monthly* and *Literary Digest* all catered to the public's interest in purchasing, operating, and maintaining automobiles. Often these articles dealt with issues such as safe driving that were largely ignored in owner's manuals. In 1918, *Good Housekeeping* promised to "answer any question about a car, its purchase, care, and use, that any perplexed motorist, actual or would-be, cares to ask," adding that "the service is free" and "all we ask for is a stamp."[40]

The queries perplexed readers posed to *Good Housekeeping* are lost, but we know it and other women's and family magazines provided the public with a great deal of advice on driving techniques. To drive well takes "years of actual experience or a combination of carefully studied written instructions and a reduced period of real experience," explained a *Good Housekeeping* article. The motorist had to be able to judge "the ability of his car, road and weather conditions, and many other factors in confusing combinations."[41] One aspect of technique taken up by such magazines was driving in winter, a practice that became more common in closed and more reliable automobiles. The article, "Keep Your Car from Catching Cold," published in the *Ladies' Home Journal* in 1920, provided a useful primer on winter operation. Readers learned that water-cooled engines were vulnerable and needed antifreeze to keep the water from freezing. "You can buy these preparations ready made at accessory stores and garages," the piece explained, "or mix your own." Assuming women would be able to follow a recipe for antifreeze as readily as for a cake, the article discussed in detail how to mix denatured alcohol, glycerin, and water to produce an inexpensive, homemade version. It also discussed how to periodically test the alcohol content of the coolant using a hydrometer, a tool designed to measure specific gravity. Readers of "Keep Your Car from Catching Cold" also gleaned other tips—the need to change to lighter oils and greases in the winter and how to keep the engine warm by covering it with a blanket after turning it off, by using a heater, or by leaving an electric light burning overnight in the engine compartment. The most important thing readers had to learn, however, was how to drive on roads made slick by ice and snow.[42] "Winter driving?" asked a *Saturday Evening Post* writer: "That, as they say in New York, is something else again." He went on to describe his first skidding experience, recalling that awful feeling when the car turned ninety degrees from the direction he was heading, threatening a disastrous crash. There was no surefire prescription for avoiding skids, he explained, as such knowledge "comes largely through experience."[43] Yet such articles both apprised readers of the heightened hazards and risks of winter operation and passed on tips that were hard to find elsewhere in print.[44]

By the late 1910s and 1920s, motorists were not only running their cars throughout the year but also ranging farther from home on vacations, often traversing the Rockies like the pioneers of old. Articles began to appear on "mountain motoring," as one *Saturday Evening Post* writer called the techniques of managing an automobile on the long, steep grades of the American West.[45] In the same magazine, another author, Courtney Ryley Cooper, wrote a particularly gory article offering advice on "how to avoid being killed while on your vacation." A resident of a small town in the Colorado mountains, he claimed to be writing as a "favor" to physician friends who disliked seeing the "landscape muddled up with the wrecks of humans and automobiles." Cooper assumed his readers lived in cities, "where the streets have been carefully graded" and where "a hill of more than 6 per cent is unusual," and knew little about mountains or mountain roads. "In the Rockies," he explained, grades of 10 percent continued "for miles," causing problems for the "flatland driver." Climbing the mountains he "begins to fight his engine" keeping the car in high gear too long and causing the engine to overheat and the radiator to boil over.[46] But it was on descents, as when coming down a mountain, that the flatland motorist most likely found himself or herself engaged in a life-and-death struggle for control of the vehicle.

City drivers might fear the "motorcycle cop," Cooper observed, but in the mountains they should fear the "speed cop who travels unseen, but who is at your elbow, nevertheless, whenever the desire seizes you to forget caution or to step on 'er. . . . His name is Death." Cooper offered some tips on outsmarting "Death" while in the Rockies. "The first two things to remember" were caution and slowness. Drivers should keep their vehicles in low gear, utilizing the engine's compression to slow down the car on downgrades. Drivers should never use the brakes alone, he warned, as they will overheat, "catch fire, become glazed, and then—all in an instant give way." It is at that moment that the "speed cop" takes over, a moment Cooper describes all too graphically:

> The machine shoots forward with suddenly doubled speed. A push at the gear lever only brings a horrible burring and a snapping of teeth, nothing more. So you do one of two things: go into the bank and smash the front end of your machine, or waltz gaily over the edge of the road to another Great Divide, from which no one ever has been able to send back a report on the scenery.[47]

As if this lurid description did not adequately make his point, Cooper added a grizzly epilogue for emphasis. The previous summer he had been following a tourist's automobile down a mountain, and the novice's car was behaving erratically. When the rattled driver pulled off the road at a flat spot, Cooper pulled over behind him. Offering to drive the man's car the rest of the way down the mountain, the motorist "attempted a laugh" and replied, "I guess I can make it all right." Querying him then about his brakes, the man replied that they were "fine." Cooper then asked whether the man had also "been using compression" to slow down his car, to which the tourist responded, "Huh?" Cooper explained how in low gear the engine's compression slowed the car and thus saved the brakes. "Oh, yes—I understand," said the man, who then drove off with his wife and kids. "Ten minutes later we heard a crash," he wrote. "The man's car had gone over the edge, somersaulted four times and crushed the whole family." Cooper's final advice: if you lacked a proper vehicle, spare tires and other supplies; if you didn't have the knowledge how to "examine your engine, your brakes, your oil and cooling system," along with "a head that can assimilate simple rules and follow them—Don't drive in the mountains!"[48]

While women's and family magazines provided driving tips and life-saving advice, they largely ignored under-the-hood tasks. This was not the case with mass-circulation magazines like *Popular Science, Popular Mechanics,* and *Mechanix Illustrated* that catered mostly to male readers interested in tools, in working with their hands, and in new technologies. In the 1910s and 1920s, as ownership of cars became more widely diffused, these periodicals began to publish short articles about maintenance and repair. Some of the most innovative appeared in *Popular Science* as part of a series titled *The Model Garage* that began in 1925. The articles blended automotive technical knowledge with entertainment and proved popular enough that they appeared every month for nearly forty-five years.

As described by the *Popular Science* editors, *The Model Garage* comprised a "fascinating new series of stories in which two veteran auto men tell you how to save worry and expense" in dealing with an automobile.[49] The protagonists included Gus Wilson, a veteran mechanic who had "been working on automobiles since the days they were called horseless carriages." Wilson owned and ran the Model Garage with his partner, Joe Clark, the "figure man" of the business. The garage was located not far from "the city," on the outskirts of a small town, large enough to have a lumberyard and a few all-night gas stations, along with a diner and a hotel, the Park House, but small enough that Gus seemed to know everybody. Although the stories about Gus, Joe, and the Model Garage were fictional, they were allegedly based on real-life situations—"the Model Garage is located in a town not far from New York City"— and many readers were convinced that the stories described their local mechanic or service station, though neither the garage nor Gus and Joe were real.[50] Written by Martin Bunn, the nom de plume for a cadre of writers who produced the series over the years, each story was very short, its triple columns of text filling at most two pages, including the opening illustration of Gus in greasy coveralls listening to a client, staring under the hood at a balky engine, or triumphantly holding up the guilty part responsible for somebody's car trouble.

The episode "Gus Tells How to Adjust a Carburetor and Shows How to Cure Starting Trouble" typifies how the *Model Garage* stories simultaneously entertained readers and instructed them about automotive technology. The minimal plot is sketched in just enough detail to draw readers in and to set up a "motor mystery," as a character in another story refers to car trouble, a mystery that Gus always solves. In this instance the setup is a morning fishing trip that Gus and Joe have taken, described in four brief sentences ending, "Thanks to Joe's skill, they had a string of speckled beauties." As the two men, heading home, "were about to turn into the state highway, Gus slammed on the brakes," the narration continues, because "stalled squarely across the trail was a mud-covered, battered touring car." Seeing nobody around it, Gus honks his horn, at which point a "small, thin individual crawled out from under the car." Gus asks what is wrong because "a stalled automobile was as much of a challenge to the veteran auto mechanic as a red flag is to a bull." The narrative proceeds:

> "Blamed if I know," replied the lanky individual. "If it was a horse, now, I could tell you, but these gasoline animals are a plum mystery to me. Leastways that goes for why it stopped. . . ."
>
> "Let's have a look at it," said Gus. He climbed out of his car and proceeded systematically to eliminate one possible trouble after another. In a few minutes he gave a grunt of satisfaction.
>
> "Here's part of your trouble," he stated. "This fool gasoline saver has come loose and air is leaking into the manifold so fast that it spoils the mixture. Here—give me a wrench and I'll tighten it up for you. When you strike the next town I'd suggest that you throw it away and put a plug in the hole."
>
> The owner grinned sheepishly. "Reckon that's one on me, stranger," he said. "The garage man in the town I just passed sold me that. He claimed that it would make the gasoline last twice as long."
>
> "Applesauce!" snorted Gus.

Following Gus's continued tinkering in the engine compartment and further interaction with the stranded stranger, the story then imparts to readers a lesson about gasoline. Observing that the gas dripping from the carburetor smelled "fierce," Gus ask the man if he bought it from the same fellow who sold him the worthless gas saver. " 'Yes,' replied the owner. 'I bought five gallons there. I got a bargain, too—two cents cheaper than the last.' " His intuition confirmed, Gus proclaims the gasoline "rotten" because the lighter hydrocarbons necessary for it to easily burn in the cylinder have evaporated. After drawing some fresh gas from his own tank and putting it into the stranger's carburetor, the engine immediately comes to life.

"I'll be durn thankful if you'll show me how to set the carburetor," the stranger says, now that his car is running, an entrée for Gus to impart additional advice. As the man profusely thanks him, Gus and Joe climb into their car and resume their return to the Model Garage. Once on the road, Joe asks Gus whether the motorist would ever "know how to take care of a car," setting up the story's final line: " 'I wouldn't bank on it,' grinned Gus. 'He'll probably end up by putting oats in the gasoline tank and saying 'Whoa!' when he ought to put on the brake.' "[51]

For *Popular Science's* male working-class readership, this story provided a quick and easy education about specious gas-saving devices, about "rotten" gasoline, and about the workings of carburetors, all topics virtually ignored in owner's manuals. That other *Model Garage* stories imparted useful knowledge for taking care of automobiles is evident in their descriptive titles: "When Carbon Clogs Your Motor, Gus Tells How to Burn and Scrape It Out" (1926); "How to Paint Your Own Car: Gus Tells Why It Pays to Be Fussy about Dust and Your Brush" (1926); "Some Fine Points of Valve Grinding: Gus Explains a Few Tricks to Motorist Who 'Knew How,' but Ruined an Easy Job" (1927); and "Stop the Noises in Your Car, Gus Tells How to Trace and Cure the Squeaks, Rattles, Hum and Thumps That Warn the Driver Something's Amiss" (1927).[52] For readers, learning from Gus would have been like learning from a wise and genial elder relative, as the unassuming mechanic wore his knowledge lightly, explaining but never lecturing. Like many of the readers of *Popular Science*, Gus's learning came not from schooling and highfalutin degrees but from years of hands-on experience. Even when Gus gently criticized or made fun of a motorist, as he does in the carburetor tale, implying that the stranded driver was an ignorant hayseed as likely to shout "Whoa" instead of hitting the brakes, readers would not feel they were being personally criticized or ridiculed. Even if they had not known the particular lesson Gus was teaching in the story—that gasoline could lose its volatility and go bad, say—they could still chuckle at and feel superior to the fictional character whose ignorance got him in difficulty. Similarly, readers were not likely to be put off by the many other negative stereotypes that appeared in *Model Garage* stories, such as thieving gypsies, penny-pinching tightwads, effete artists, pedantic professionals, and most notably, gabby and unmechanical women.[53]

Along with playing to widely shared stereotypes, the Gus stories entertained by capitalizing on the immensely popular detective story genre. In most every episode Gus confronts a "motor mystery" and readers are kept in suspense as he gathers evidence about the case before offering up a solution. Gus's mastery of deductive reasoning sometimes prompted comparisons with Sherlock Holmes, the famous fictional detective created by Arthur Conan Doyle. In a 1938 *Model Garage* story, "Gus Tunes a Car by Ear," the master has claimed that car troubles can sometimes be diagnosed without even looking at the vehicle, but Gus's young helper at the garage, Harry, does not believe his boss. The two make a bet on the proposition, and soon afterward a woman drives into the garage and complains to Harry that her car's engine isn't running smoothly. Harry seeks out Gus in the office and suggests he take a look at the car, but Gus had heard the car drive in and has already solved this mystery:

> "I don't need to take a look at it," he said. "You take the distributor head off. You'll find a Y-shaped crack in it. Drill a hole right where the Y branches. Make the hole twice as big as the width of the crack. I'll be out in a few minutes."
> "All right," Harry said. "But it sounds screwy to me."

Screwy or not, when Harry removes the distributor cap there is the Y-shaped crack Gus had predicted; when he drills the hole as instructed, amazingly the motor's erratic behavior suddenly vanishes; Gus explains that the fix is only temporary, but it gives him time to procure a new replacement cap. The next time Harry sees Gus alone, he exclaims, "All right Sherlock Holmes, let's have it," setting up the veteran mechanic's, "elementary, my dear Watson" moment as he proceeds to explain to his dazzled subordinate (and of course to curious readers) how small cracks in the distributor cap can cause irregular firing because electricity flows along those faults rather than to the spark plugs.[54]

Over time, *The Model Garage* fiction followed the same developmental pattern as owner's manuals; that is, it evolved from a technical, do-it-yourself focus toward one emphasizing entertainment and consumption. The Gus stories from the 1920s implied a readership that still might have been maintaining their vehicles at home, although stories about burning carbon out of cylinders, grinding valves, or painting one's car already in the 1920s described tasks that probably few readers were likely to take on themselves. By the late 1930s, the Gus stories, if they dealt with tinkering at all, addressed minor tasks beneath the hood, as in "Spotting Ignition Trouble," say, or "Trouble-Hunting When Your Car Won't Run," both from 1936.[55] Increasingly, however, the series emphasis shifted, as titles like "Can You Afford a New Car?" "Are You Wasting Gasoline?" or "Keep Your Car Looking New," also all from 1936, spoke to readers as consumers, driver-possessors who have their automobiles professionally serviced but who enjoy reading about them, not tinkering with them.[56]

Because every car owner is also a former car buyer, it is not surprising that *Model Garage* episodes periodically addressed the issue of choosing and shopping for a new one.[57] In an early story from 1926, "How to Pick the Best Car in Its Class," Gus and his nephew Henry, along with the latter's wife, Grace, are visiting an automobile show. The young couple are in the market for a new car and are arguing over what it should be: "I'd like to get one of those sporty-looking roadster types," says Henry on noticing one vehicle, prompting his wife to respond, "Why, Henry, you promised a sedan." The exchange elicits the following from the sage old mechanic: " 'Grace is right, son,' Gus said. 'If you can have only one car, the best buy is a closed model, but I'd recommend a coach," meaning a two-door sedan. They discuss other issues including color before the subject turns to engines. "What's the dope on overhead valves, uncle?" asks Henry of Gus. "Are they really much better?" "Theoretically they are—all the racing cars use them," Gus replies, continuing without satisfying his nephew.

"Humph!" exclaimed Henry. "It sure is a hard job to pin you down to a definite opinion as to what is really the best." Gus smiled.

"For the darn good reason, son, that there isn't any such thing as best. The fact that there are so many different ways in which gasoline motors are made shows that the designers can't agree. Whenever there are two ways to make a piece of machinery, there are always good arguments for either way, and you'll have to decide for yourself which of these ways fits in best with your way of doing things."

Pushing Gus further, Henry hastily retorts: "Do you think a car with four-wheel brakes is safer, uncle?"—this merely elicits another qualified response from the veteran and leaves Henry more frustrated than ever. In 1926 it was perhaps technically defensible to refuse to pronounce definitively on the merits of overhead versus side-valves or even four-wheel over two-wheel brakes, but Gus's refusal in another story to recommend one of the two types of passenger compartment heating devices—manifold heated air versus a separate gasoline-fueled heater—seems perverse. "We don't play favorites," Gus laconically explains to an inquiring customer, refusing to back the less dangerous and soon standard manifold heated approach to keeping the car warm and by implication ignoring the known hazards of gasoline heaters.[58]

GUS'S REFUSAL TO "PLAY FAVORITES" in the *Model Garage* stories most likely stemmed from the unwillingness of the magazine's editors to publish any content that might offend any of its advertisers. Never in the stories does Gus give readers an impartial, expert opinion on, say, the strengths and weaknesses of the Ford versus the Chevrolet among low-priced cars, or the Buick versus the Chrysler in the higher-priced bracket. Gus never reveals what *he* thinks to be "the best car in its class." This same reluctance to play favorites, to name names and critically discuss or analyze automobiles characterized virtually all of the many books, magazine articles, and newspaper columns dealing with motor vehicles published in the United States before World War Two. This situation perpetuated one of the dilemmas of technology consumption: How could a technically untrained individual make an informed choice among the many types, makes, and models of vehicles (or any other technology) on the market?

In the 1920s, there were indications that the journalistic taboo against advocating "favorites" among a group of products was starting to collapse, thanks to the consumer rights movement and what at the time was called "scientific purchasing." In 1927, the young economist and engineer, Stuart Chase, and his collaborator, F. J. Schlink, published *Your Money's Worth, A Study in the Waste of the Consumer's Dollar.* The book's success led the authors to found the product-testing company Consumers' Research, Inc., which as early as 1928 was informally circulating lists of endorsed products, including automobiles, to its subscribers. It hailed the new 1928 Ford Model A as a "remarkable development" but also categorized some vehicles as "not recommended." By the 1930s, its informally circulating lists had expanded into a monthly magazine, the *Consumers' Research Bulletin,* which began the practice of each year devoting an entire issue to automobiles.[59] "Automobiles of 1936," for example, consisted of eighteen pages of print and illustrations. Here for the first time readers might find critical appraisals of the sort long available about movies, books, and theatrical performances but never before available on automobiles or other consumer products.[60]

The *Bulletin's* authors divided Detroit's 1936 models into five price groups and graded each make and model in a group with an "A," "B," or "C." They also listed their first, second, and third choices in each group. In "Price Group 1—$415 to $590," for example, the reviewers awarded "A" grades to the Ford V-8 Standard, the Plymouth business sedan, and the Chevrolet Standard, ranking them first, second, and third respectively. Prospective buyers could either base their shopping decision on the car receiving the highest ranking or, should they wish to make up their own mind in choosing from among similarly graded vehicles, could read the full assessments before making a choice. The authors helped readers reach their own conclusions by providing detailed information for and considerable analysis of each car, including its potential depreciation, calculated by using "the difference between the original New York delivered price of the most nearly comparable 1935 [model] and the retail sales value on that car turned in as a used car." Consumers could also learn about factors that affected operating cost, such as the "average brake factor," a calculation that sought to measure each car's consumption of brake linings, which were components that wore out regularly and were relatively costly. From graphs showing the mileage each car delivered per quart of oil and gallon of gas, consumers could also get a sense of what they might have to pay on a day-to-day basis to run any particular car.

If readers wondered why the Ford was ranked ahead of the Plymouth and Chevrolet, even though all three cars received a grade of "A," they learned that Ford bodies were all steel, unlike those of the Plymouth, "the only car in this group using a wooden frame for the body." And "the *Stromberg* carburetor used on the *Ford* is unquestionably superior to the *Carter* used on the *Chevrolet*," the authors explained, adding that the Carter "gives considerable dissatisfaction, understood to be due to both manufacture and service difficulties. This carburetor uses a tapered metering pin which varies the opening of an orifice, thus partially controlling the mixture ratio. It is believed that this device has given *Chevrolet* service men considerable trouble." Consumers' Research experts did not simply offer their own opinions but also incorporated the perspectives of automotive professionals into their appraisals.

Whether readers of the *Bulletin* only skimmed the bottom line results or read and studied the evaluations of each automobile, they finally had the kind of impartial and knowledgeable advice consumers had always lacked before. Yet few Americans ever saw the magazine. Historian Charles McGovern, in his study of Consumers' Research, notes that one had to join the organization in order to receive the *Bulletin* and only a few thousand members of "the educated middle class" did so. Additionally, other factors limited the publication's appeal including "miniscule print and almost no illustrations"; its appearance resembled "government reports or legal briefs." But the primary reason the *Bulletin* had such minimal influence, McGovern argues, was the insistence by Consumers' Research that members not show the publication or share any of the information in it beyond their family circle. The organization believed that by restricting the circulation of the product evaluations published in the *Bulletin,* it was protecting itself against potential libel suits by corporations whose products might be criticized. The organization reasoned that if negative things were published in the *Bulletin* about cars or any other products, these statements could not be deemed "public speech" and therefore be actionable at law as long as circulation was restricted to a paying membership. But such lawsuits "never materialized" McGovern reports.[61]

In 1936, disgruntled employees of Consumers' Research launched a rival product-testing organization, Consumers Union, which then started publishing its own magazine, *Consumer Reports*. Although this magazine, too, was sold on a subscription basis, there was never any effort to restrict its readership. Its circulation grew slowly at first, but in the years after World War Two *Consumer Reports* found a growing audience among newly prosperous consumers seeking expert, objective advice on the purchase of automobiles and other products. Finally, the decades of consumer bewilderment and confusion were ending. Reading, useful and even essential in learning to use, operate, and maintain personal technologies from the beginning, increasingly became important in shopping for them as well.

Invaluable though the automobile evaluations appearing in the *Consumers' Research Bulletin* of the 1920s and *Consumer Reports* in the 1930s were, they did not give readers a firsthand account of what it was like to actually *drive* the vehicles being tested. These articles read like laboratory reports, analyzing vehicles as if they were scientific specimens. The authors were anonymous and relied excessively on the passive voice as if to hide their feelings and personalities in the interest of scientific impartiality. Readers therefore got no impression of what, say, the seats felt like or how the car rode, steered, or handled. This final piece of modern consumer product criticism fell into place immediately after World War Two through the journalistic innovation of the "road test."

The first road test appeared in the February 1946 number of *Mechanix Illustrated,* a male hobby magazine similar in content and readership to *Popular Science* and *Popular Mechanics.* Responding to the first new cars to roll off Detroit assembly lines since civilian production had been curtailed after Pearl Harbor, the piece carried the unassuming title "MI Tests the New Cars." The author was identified as "MI's auto consultant, Tom McCahill." In his debut article McCahill evaluated Ford's new two-door sedan and the Series 50 Buick. "As soon as I heard that the new Fords were coming off the production line," McCahill explained in the colloquial, personalized prose that would become his hallmark, "I rushed to Dearborn, Michigan, and after getting the go-ahead from company officials, I was given a stock car and proceeded to 'give it the works.'" He continued, shifting his prose into hyperbolic gear: "I took the two-door sedan I had selected out for a good fight, and was I surprised—the tougher I got, the better the new Ford performed." He was "impressed" by the car's

riding quality, something that never had been a feature in any Fords
of the past. At fifty miles per hour on some very bad roads I purposely

hit deep ruts and holes with little or no effect. Saying to myself, 'This can't be true, it's not Ford-like,' I spied a railroad track that crossed my path and noticed that at the road edge it stood above the pavement a good 5 or 6 inches. Without slackening speed, I hit the rail. Aside from a slight thumping noise and a bit of feeling in the steering wheel, you might have thought I had hit a small stone on the highway.

Next the brakes got "the works" and passed easily; being "larger than before," McCahill explained, they "are designed to carry nearly twice the normal brake load." Testing the new Ford's engine, also larger and reportedly more powerful than the pre-war model, "it seemed to lack some of the punch of the past," he added, leaving him "unimpressed" with the car's pickup, or acceleration. This resulted, he suggested, from an "extra piston ring" being added on each piston in the new model; this took "the Ford out of the oil-burner class," McCahill noted, using the derogatory term for diesel vehicles and others that consumed lots of oil, but the additional friction created by the extra rings moving in the cylinder "cut down the punch" of the car's performance.[62]

McCahill's first automobile road tests displayed the same interest in performance and the behind-the-wheel qualities that are characteristic of the genre today. Commenting on the 1946 Buick's acceleration, he said the car "picked up to forty or fifty in a matter of seconds, and without hesitation."[63] In another review, describing the performance of the 1946 Oldsmobile 76 in winter conditions, he observed, "When I opened the throttle wide the power instantly began to build up and I gained momentum like a rocket, in spite of the snow."[64] Testing a 1946 Packard Clipper, he observed that it "has all the pick-up, speed and hill climbing quality anyone could wish for" and, in a mixed-metaphor aside, concluded: "its power plant is the little brother of the giant Packard engines which drove our PT boats in the Pacific" so it "gathered speed as if jet-propelled."[65] After driving the 1948 Hudson, McCahill rapturously told readers how it "would whip from 10 to 70 mph in a matter of seconds." Summing up this production hot-rod's performance, he concluded: "In the six-cylinder class there isn't a car in the country that will out-perform the Hudson six, and that in the larger eight-cylinder class the Hudson eight will make an excellent showing when pitted against the best," a claim that proved prophetic, as Hudsons would soon excel on stock car racing circuits nationwide.[66]

Although McCahill enthused about cars that accelerated "like a rocket" or that felt "jet propelled," his early road tests contained no hard data in the form of elapsed times for the quarter mile or for accelerating from zero to sixty miles per hour, both now standard criteria for measuring automobile performance. Readers may have complained about this obvious shortcoming, for in the same review in which he extolled the eight-cylinder Hudson's competitive potential if "pitted against the best," he added: "I feel perhaps a word on how I make these car tests may be of interest here. On nearly all cars tested, stop-watch records are made of the acceleration in all gears. Further performance tests are usually made in high gear over a measured mile, starting at various speeds. Also, whenever possible, climb tests are made and recorded." He went on to explain that the "detailed figures" resulting from such stopwatch tests "are not given in the interest of fairness." One tested vehicle might be right off the showroom floor with little mileage while another might have "thousands of miles on it, which meant a thoroughly broken-in automobile." Because "most good stock cars are not really thoroughly broken in before they've gone 5,000 miles," he asserted, to publish the results of time trials would unfairly compare apples and oranges.[67] Whether McCahill's "interest of fairness," *Mechanix Illustrated* editors' worries about embarrassing car companies by publishing unflattering data, or his personal relationship with Detroit officials was really lay behind the withholding of stopwatch data is not clear. At some point in the 1950s McCahill or the editors reversed position, and *Mechanix Illustrated* along with other automotive periodicals thereafter consistently published the concrete acceleration data about vehicles in road tests.[68]

McCahill also pioneered the now ubiquitous practice whereby critics register their subjective impression of what it felt like to sit in and drive a particular car. He commented on design features such as the layout of the driver's compartment, seat comfort, and steering wheel position, all of which contributed to enjoyment and safety behind the wheel. Sometimes photographic illustrations conveyed these subjective impressions in McCahill's road tests. For unlike illustrations in earlier automotive periodicals, which never showed a human being with a vehicle, McCahill himself often was in the picture. Seeing him standing next to the 1948 Hudson with his arm held straight out from his body and *over* the car's roof made visually evident the car's startlingly low overall height. In other road tests, photographs taken either from the front passenger seat or over McCahill's shoulder gave readers an impression of the driver's-eye view of the steering wheel and controls. A photograph showing him giving the new Ford "the works" in his debut review from 1947 was taken from a distance as he sped through a "water test" demonstrating the car's leakproof qualities. All of these photographs enabled readers, in a phrase coined by historian David Nye in a different context, to "imaginatively possess" the vehicle, to project themselves into the driver's seat and, as a kind of thought experiment, test-drive it themselves.[69]

In the second half of the twentieth century, the road test genre spread to an even wider range of products as its value to technology shoppers remained unrivaled. Consumers wanting to buy new and bewildering machines like digital cameras or computers benefit immensely from such articles.[70] They now can find performance reviews of such devices not only in magazines and newspapers but also online. *The New York Times* regularly publishes columnists who in effect road-test the latest personal technologies and report on their experiences. A few years back the paper ran a piece, "New PC: How to Kick the Tires," offering readers tips as to how to conduct their own tests when shopping for computers. The borrowing of metaphors and expressions from our collective behind-the-wheel experience evident in the article's title is characteristic of the computer age. "Test Drive a Macintosh," exhorted a 1990s Apple Computer advertisement, while a PC Brands ad of the same era asserted that "while the other guys are blowing smoke, PC Brand is burning rubber."[71] People have long been speaking of an "information highway"; some claim "the PC has become the dashboard of our lives."[72]

Given how difficult and intimidating consumers found the jargon connected with automobiles ninety or a hundred years ago, it is ironic to see it enlisted to make digital computing seem familiar and friendly. As anyone who has owned and used personal computers well knows, living with this complicated technology is hardly like kicking the tires or glancing at the dashboard of a car. Our struggle with computers is not unlike ours long ago with automobiles; it is a story we are still living, and one to which we turn in the final chapter of this book and its epilogue.

Chapter 6

Computers and the Tyranny of Technology Consumption

I N 1984 CRICKET TOWNSEND, the president of a personal computer user group in Fremont, California, was angry. What was "the first thing you learned about your computer?" she asked the readers of the group's newsletter. Was it, she prompted, that

> there should have been a sign that stated, "BUY AT YOUR OWN RISK"? HOW MANY GOOD PROGRAM DISKS HAVE YOU ERASED? HOW MANY GOOD MANUALS HAVE YOU TORN TO BITS BECAUSE THEY WERE OF NO USE? HOW MUCH OF YOUR HAIR HAVE YOU PULLED OUT BECAUSE SOMETHING WOULDN'T WORK? HOW MANY TIMES HAVE YOU CUSSED A BLUE STREAK?

When Townsend unleashed this diatribe, the personal computer was not yet a decade old. "If you thought I was talking about you in the above statement and only you," she continued, "you're wrong. If you could see into other members' homes as they read the above you would see the expression on their faces that said 'SHE'S TALKING ABOUT ME.' The point is this," she concluded. "Everyone goes through the same (pardon my French) HELL. You are NOT the first and will never be the last to feel these feelings."[1]

The last time consumers experienced such hell with a new technology had been with the arrival of automobiles earlier in the century. Americans paid a heavy price for their new mobility. They were bewildered as first-time buyers choosing one automobile from the dizzying number of types, models, and makes available. They faced hell again when they first got into the driver's seat as utter novices and attempted to start, steer, shift gears, and otherwise manage their new machines. They also spent hours breaking fingernails and skinning knuckles while tinkering under the hood when "something wouldn't work." And when they sought help and advice from silent instructors or owner's manuals, they found them opaque, exasperatingly incomplete, or "of no use" at all.

By the time personal computers came on the market in the latter 1970s, Americans were too young to remember early automobiles and motoring. Over the years, they had eagerly adopted other consumer technologies: electric stoves, refrigerators, washing machines, power drills, vacuum cleaners, phonographs, steam irons, radio and television sets, to mention but a few. While not everybody possessed all of these machines, and access to any of them depended on income, surveys reveal that especially in the years after World War Two even the poorest households acquired a variety of mechanical, electrical, and electronic devices.[2] Indeed, the unprecedented growth in consumer spending in the postwar years depended in large part on people filling their homes and apartments with personal technologies.

None of these machines subjected consumers to the kinds of hell cars and computers did. Learning to operate them required little new knowledge or the acquisition of new skills. Consumers already knew how to plug in a new electric drill, popcorn popper, waffle iron, or vacuum cleaner and then how to flip its switch (or manipulate a lever, button, or trigger) to make it work. Managing the on-off interface of such appliances was so simple and easy that people seldom felt compelled to read the manual or instructions. People did not speak of "operating" a popcorn popper or microwave or think of themselves as "operators"; the work attendant to using such technologies is too easy or trivial to warrant such terminology. Some of these machines, refrigerators for instance, did not even have an on-off switch and started running once plugged in to a power outlet. Others, such as cake mixers, blenders, or toasters, may have had an additional control or two enabling users to adjust their speed, temperature, or running time, but consumers were seldom puzzled by such minor complications. Indeed, many appliances, such as electric mixers or power drills, simply mechanized tasks once performed with hand tools. Using an electric cake mixer, cooks combined eggs and flour into batter essentially the same way they previously had using a fork or hand-operated eggbeater; woodworkers bored holes with power drills pretty much the same as they had for centuries with brace and bit. Machines were faster and required less muscle power, but the skills of baking or carpentry were not centered on the mastery of machinery; they lay in the knowledge of the qualities and behavior of eggs, flour, or wood; in choosing appropriate recipes or plans; and then following them step by step in proper sequence while creatively responding to the inevitable little problems that occurred in the course of the work.

Even the electronic entertainment technologies that began with radio in the 1920s and which included hi-fi and stereo systems, and most notably television in the 1940s and 1950s, caused relatively little frustration among users. To be sure, *choosing* a particular model and brand among competing devices could be confusing, as shoppers had little way of assessing the often baffling technical claims made regarding different machines. But *using* the devices required little new learning and less skill. A mass market for wireless, or radio, followed quickly after the first broadcast in 1920, as commercially manufactured radio "sets," or receivers, came on the market. The earliest adopters of radios did have to learn how to manage the wet cell batteries that powered their receivers, for radios did not yet operate off household current (which at the time many homes lacked anyway). This involved regularly testing the acidity of the battery's electrolyte and replenishing fluid when necessary. Tuning those early sets was more complicated than it later became, as they had two, separate tuning circuits, each with its own knob or control, which had to be worked in tandem. Nevertheless, millions of people bought these early radios and adapted to the challenges of operating them without complaint. Within the decade technological change rendered the messy batteries obsolete as new radios became "servants of your light socket," as one magazine put it, noting that a "radio receiver powered directly from the light socket is like the automobile with a self-starter. Pushing the button starts the machine."[3] Tuning, too, was simplified as the one-knob or single-button procedure we know today came into being.

Television, which debuted at the end of the 1930s and reached the marketplace in quantity after the Second World War, replicated the relative user-friendliness of early radio. Early adopters had even less to do in tuning their TV sets to a particular channel than had early radio users, for the channel controls were preset for the mere dozen or so available stations.

Getting a decent picture, however, on those early black-and-white sets was another matter because the reception provided by rooftop antennae, or, more commonly the rabbit ears that sat atop the set itself, varied greatly. Because TV signals moved in a straight, line-of-sight fashion, reception was distorted or blocked altogether by hills, buildings, and other environmental obstacles. So after selecting their channel, users had to fiddle with the controls regulating vertical hold, picture density, lightness and darkness and then fine-tune the position of their rabbit ears to turn what might be a snowstorm on the screen into a recognizable picture. For decades consumers tolerated often spotty reception, until cable all but eliminated problems of signal interference, and until digital, computerized HDTV, appearing at the end of the 1990s, once again complicated the user's life—but that is to get ahead of our story.

Consumers also adapted relatively easily in the middle decades of the twentieth century to devices that recorded sound and images. New kinds of cameras, such as single-lens reflexes using 35 millimeter film along with movie cameras and Polaroids sold well in the postwar years. All of these technologies demanded a bit more than just pressing a button as with the original Kodak snapshot camera of the 1880s. But even beginners quickly learned to take passable photographs and shoot entertaining home movies. Similarly, consumers who bought the new audio recording devices that came on the market, at first using wire as the recording medium followed by machines that used magnetic tape, learned to handle the delicate material, thread it through the innards of their machine, and manipulate the various cranks, buttons, and knobs required to make the recorders work. Millions used such machines to document their child's music recitals or other quotidian activities of life and left little evidence that, after getting the hang of things, they found these personal technologies particularly demanding or frustrating.

All of these post-automobile technologies also were relatively easy to maintain. Purchasers were not expected to use tools and tinker, as had been the case with early cars, but they did have to satisfy their machines' hearty appetites for supplies: film for cameras, tape or records for sound recorders or phonographs, and vacuum tubes for radios. Vacuum tubes, like light bulbs, occasionally burned out, and consumers, though they might know nothing about electronics, could replace tubes themselves. By the 1930s and 1940s, drug and hardware stores often had do-it-yourself testing machines and stocked popular sizes of radio tubes. Customers would bring in a suspect tube, insert it into the machine for the test, and then if the tube was confirmed to be bad consult a chart to select and purchase an appropriate replacement.[4] During the 1950s, vacuum tubes gave way to transistors, making electronic devices more reliable; at the same time transistorization ended ordinary consumer activity under the hood, so to speak, of their radios or television sets as people now had to turn to professionals for any servicing. Buyers were neither expected to diagnose the reason electronic devices failed to work nor to fix them. Fortunately, most of the time such machines and appliances performed without trouble. If something did go wrong, however, and consumers consulted the instructions that came with any piece of electronic equipment, they rarely learned anything about making repairs. They might find a list troubleshooting tips to check if their dishwasher, television set, camera, or vacuum cleaner failed to perform, but for anything beyond an obvious and easily rectified problem, they were advised to turn to an authorized repair facility. In general, after automobiles became user-friendly fixtures of daily life by the late 1930s, the new technologies that followed caused consumers little struggle and frustration. In the late 1970s and early 1980s, this relatively idyllic relationship consumers had with new machines ended abruptly, shattered by the arrival of personal computers and the digital era.

The hell experienced by early adopters of personal computers may be glimpsed in newsletters such as the one in which Cricket Townsend, quoted at the beginning of this chapter, voiced her rage. Published by the Society for the Prevention of Cruelty to Apples (that is, Apple computers) in Freemont, California, the newsletter was one of many put out by thousands of computer user groups that sprang up around the country, organized by people needing help with and wanting to help others by sharing information about the new technology. We can also track the struggle of early users in articles, letters, and product reviews that appeared in computer magazines during the late 1970s and 1980s, such publications providing help and advice to beginners and other users just as motoring periodicals had in the early 1900s. And like many adults who are middle-aged or older, I myself experienced the travails of early adoption, having purchased my first computer in 1982. Today I am finishing this book on what, if I count correctly, is my eighth computer, but I still vividly recall the early troubles I suffered (and sometimes today still endure) adapting to the ways of this complicated technology. Indeed, as noted in this book's introduction, it was a computer-related problem of my own some years ago that suggested to me the uniqueness of technology consumption and its potential as a topic for study. In the remainder of this chapter, I will compare the early struggles of computer adopters with those encountered by the first generation of automobile users, allowing us to see more clearly both the differences and similarities in the user unfriendliness of electronic and digital technologies versus earlier mechanical and electrical ones. In this account I shall draw on published newsletters and periodicals as well as on personal recollections.

UNLIKE MOST OF THE DOMESTIC TECHNOLOGIES we have been discussing, the computer began as a machine exclusively owned by large institutions; its development as a personal tool came later. Even today, most of the computers in offices and business settings are not "personal" in the sense we have been using the term; that is, users neither bought them with their own money nor may they take them home to use for their own or family purposes. Among those of us who *do* own computers, probably most of us use them both for personal as well as professional or occupational purposes, and we are responsible for buying, maintaining, and eventually replacing them. Indeed, the personal computer has been a major reason for the much lamented blurring of lines between work and leisure, for our ability to work both at home and in the office—or almost anywhere—and to do so any or all of the time.

Yet when electronic digital computers were first invented, nobody dreamed they would ever become personal at all. In 1943, scientists, mathematicians, engineers, and technical staff at the University of Pennsylvania, working secretly under a contract from the U.S. Army, began building what would be the first digital computer in the United States. ENIAC, standing for "Electronic Numerical Integrator And Computer," was not completed in time to aid the war effort, but the room-sized machine with its thousands of vacuum tubes, along with similar "giant brains," became an essential weapon in the Cold War.[5] Defense projects like the SAGE early missile warning system and NASA's satellite and manned space efforts in the 1950s and 1960s all depended on such computers. During the same postwar decades, government agencies like the Census Bureau and Social Security Administration and corporations reliant on massive quantities of statistical data like life insurance companies also acquired computers that now were being manufactured by firms such as IBM, a major producer of business and office machines.

How those business computers and giant brains gave rise in the latter half of the 1970s to the first personal computers is a fascinating and oft told story, one of technical innovation and miniaturization featuring counterculture idealists and computer-savvy entrepreneurs.[6] None of these early personal computers, however, had much appeal to ordinary consumers. Computer historians consider the first of them to have been the Altair 8800, a $400 kit for a computer that buyers built following circuit diagrams and using soldering irons. Advertised in the January 1975 issue of *Popular Electronics,* the Altair offered, as historian Paul Ceruzzi has observed, "a way for thousands of people to bootstrap their way into the computer age."[7] Given that the Altair came neither with software (you programmed it yourself by setting a maze of switches on its front panel) nor any output device, such as a printer or even a monitor whereby one could see the results of one's programming, only electronic hobbyists and computer buffs bought one.[8] By 1977, the Altair was joined on the market by personal computers that did not require assembly and which could be used off the shelf. These included Radio Shack's Tandy TRS-80, the Commodore PET (standing for "Personal Electronic Transactor"), and the Apple II. The latter machine, the first offering from a start-up called, Apple Computer, has been called the first "real consumer product" of the infant industry.[9] Yet it and the two other products still lacked much consumer appeal. All three used a particularly finicky means to store data—tape cassettes—and "it took VERY careful adjustment of the volume and tone controls on the cassette recorder to get programs or data to successfully load," one Apple II user recalled; another claimed that only in mid-1978, when Apple added a floppy disk drive to its machine, was it "transformed from a gadget only hard-core hobbyists would want to something all sorts of people could use."[10]

Perhaps "all sorts of people" *could* use a personal computer in 1978 but very few had any desire to do so. Indeed, many had no idea what a computer was or what one looked like, and there was still very little people could do with these machines besides play games or write computer programs (most computers came with a version of BASIC, a simple programming language). Thus, the technology had little resonance with the larger public and was of interest primarily to hobbyists. By 1980, this was changing as newly devised software began to expand the utility of personal computers, most notably two so-called "killer apps"—applications that were themselves attractive enough to convince people that they wanted to buy a computer just to be able to run the program. One was VisiCalc, the first electronic spreadsheet, which appeared in 1979. Written for the Apple II, it soon was adapted for other machines and started to make computers attractive to small businesses.[11] About the same time word processing programs like WordStar came out for personal computers, making them into writing machines.[12] For those accustomed to writing with a typewriter, the computer promised the ability of easily correcting and changing what one typed; its seduction for humanists in the academic world like myself was as a kind of super typewriter, and many of us would not use a computer for anything else for a decade or more.

By 1984, word processing had become the most common use to which people put personal computers. In a poll that year, 67 percent of those with computers at home identified their word processing program as the one they most used while 85 percent of those who employed computers primarily in their business identified it as the paramount application.[13] Still, for consumers who did not have a business or do a great deal of writing, a personal computer could seem an expensive and not very useful machine. To many people the purpose and function of the new technology beyond typing seemed opaque and hard to fathom; it was unlike the automobile in the early twentieth century, whose utility for rapid, personal travel had been readily graspable. As an Apple Computer manager somewhat awkwardly put it, consumers considered the "end use utility" of personal computers to be "unknown."[14] Not surprisingly, as late as 1980 only 377,000 were sold, although that number nearly doubled in 1981 to 646,000 units and then doubled again in 1982 to 1.3 million.[15] Slowly but inexorably, the demand for personal computers was growing, although it would take eighteen more years, until 2000 according to the U.S. Census Bureau, before half of American households possessed one.[16]

In its January 3, 1983, issue, *Time* magazine honored the personal computer as "machine of the year" for 1982 and illustrated one on its cover. This departed from the magazine's previous practice of honoring a man of the year (later a "person of the year") and registered its belief that the computer was here to stay. In a related story later in 1983, the veteran *Time* writer Otto Friedrich—who himself never abandoned his ancient, manual Royal typewriter—observed that "by the millions" the personal computer "is beeping its way into offices, schools, and homes." While predicting that "the most visible aspect of the computer revolution, the video game, is its least significant," Friedrich went on to note how the technology was already "changing the way" ordinary men and women worked at home and in the office. "So the revolution has begun," he concluded, "and as usually happens with revolutions, nobody can agree on where it is going."[17]

One direction it was going, painfully realized by the millions of men and women who acquired beeping machines, was straight to hell, at least in the sense that the consumer's struggle with the new technology replicated that which accompanied the introduction of automobiles earlier in the century. Early adopters struggled as shoppers, having to choose one particular machine and system from a bewildering number of possibilities; they then struggled to learn the arcane procedures required to use their computers; and when something didn't work, they struggled to figure out what was wrong and get their machine to behave. Finally, if they sought help from the silent instructors, or "documentation," that came with their equipment, they struggled with cryptic, jargon-laden prose. Let's look at each of these four paradigmatic activities—shopping, operating, tinkering, and reading—and compare ordinary people's early computer experiences with those who struggled with early motor vehicles many decades before.

Although the personal computer industry was less than a decade old in the early 1980s, individuals who decided they wanted to buy had much less difficulty finding machines to look at than did pioneer automobilists before 1905. Already in 1984, 50 percent of those who acquired machines for home use bought their computers through retailers (36 percent purchased by mail order, 10 percent from wholesalers or manufacturers, and 4 percent from mass merchandisers like Sears).[18] Consumers bought Tandy TRS-80 machines at the thousands of Radio Shack stores, and a variety of brands, including the recently debuted, IBM-PC, were available at Computerland outlets, a growing chain retailer; department stores such as Macy's also set up electronics departments to sell computers.[19]

Prospective computer buyers sometimes encountered a problem not experienced by early automobile shoppers. "I'm at one of the local Apple stores in my area," complained a disgruntled man in a letter to the manufacturer, and the "sales representative is playing a game on the Apple with someone from the store. I try to get the salesman's attention once, twice, and again but alas it's to no avail. The salesman and his friend are much more interested in playing the game than in me, a customer."[20] He might well have complained about children, too, for computer stores appealed to kids as no automobile dealership ever had. Faced with hoards of children playing with his computers and intimidating adult customers, the manager of a Computerland store in Los Altos, California, came up with a novel solution. He sent each kid home with a letter that began,

> Dear Parents,
> Your child has expressed an interest in working with the personal computers that we have available for demonstration at Computerland. We welcome our younger customers because of their fresh ideas, enthusiasm and creativity. Unfortunately, there are times when a few children become a lot of children for the number of computers that are available. As their enthusiasm increases, it becomes disruptive to our customers and the children are asked to leave. To the child that has just arrived, this is not fair. Often he will justifiably complain (after all, he or she just got there) but will still be asked to leave.

Wishing "to be fair to our younger customers," he asked parents for "support and assistance" and included with his letter "a sample of a card which will give our customers under the age of eighteen 15 minutes of computer time per day, per month. The card is free!!" But to get a card for their child, parents had to personally pick it up at the store.[21]

Shopping for a computer probably generated more anxiety among first-time buyers than had been the case earlier shopping for an automobile. "Computers are by their very nature intimidating," wrote Erik Sandberg-Diment the first computer columnist for the *New York Times*, in 1982, acknowledging that he too suffered from "Computer Fear Syndrome." For "the first few times in a store, I too looked, but dared not touch," he explained, but children "seem to know no fear."[22]

What made computers intimidating to adults was their abstractness. The opaque exteriors of these "black box" technologies revealed nothing about how they worked or what they did; unless a machine was turned on and running, one got not the slightest clue looking at it what its potential might be. With an automobile in the early 1900s, a prospective purchaser could take a ride with a demonstrator and, on being whisked up and down a hill at a fast clip of fifteen miles an hour, say, grasp the machine's capability and have a good sense of its usefulness. Prospective computer buyers, however, found it impossible to gain any similar understanding while watching somebody at a keyboard.

The biggest challenge for first-time computer shoppers was deciding on a particular machine and its accessories. By the early 1980s, many brands of personal computers were on the market and, like early automobiles, not all machines did the same things and most were not really usable without the purchase of extras, such as software and a printer. Buyers faced many choices, all of which immersed them in confusing technical jargon. "All this talk of 'bites . . . Kay's (who' she) . . . mega. . . floppy . . . Z-cards . . . DOS . . . peripherals,'" recalled one overwhelmed novice who then realized: "Gosh! This is a whole language in itself, and I did not speak one word of it."[23] For beginners, understanding computerese was more difficult than "Motor," not only because the language included new terms like *software* and *modem* plus unfamiliar acronyms like RAM, ROM, and DOS but also because it gave puzzling new meanings to old and familiar words and phrases, ones like *open, boot, go to, save, drive, processor,* and *memory*. Long considered a uniquely human attribute, the power of recall now was confusingly attached to a machine. "Encounter a computer salesman," Sandberg-Diment wrote in 1982, "and probably the first thing he will mention in his spiel is memory. 'Now this dandy little model over here comes with 48K of memory, and that deluxe model over there has 64K.'" He suggested that "understanding the concepts involved in computer memory should be among the first chores" shoppers took on, "but it isn't easy," he warned, "and it requires a certain amount of concentration and a willingness to forgive jargon." First-time buyers often get a computer "with too little memory to meet their near-future needs," he added, "and they always regret it later."[24] But how could buyers decide how much memory they needed, or would need in the future, without any experience with or understanding of computers?

The impossible dilemma of choosing among many versions of an often incomprehensible technology was humorously satirized in an 1980 cartoon in *The New Yorker*. A middle-aged man has walked into the "Computer Center" and, standing amid the different machines offered for sale, asks the young salesman: "Do you have one that can help me decide which is the one for me?"[25] While machines were obviously incapable of making such decisions, the desire to delegate the choice to somebody else—even a machine—was understandable. Some early adopters, especially those who worked for large organizations, often avoided the dilemmas of decision making because others chose their hardware and software for them. In the autumn of 1982, as a faculty member at Stanford University, I got my first personal computer through such a process, saving me from many agonizing and difficult technical choices. The school had subjected three different computer, software, and printer packages to a rigorous trial the previous summer and then chose the one that it subsequently offered to faculty at a heavily subsidized discount. Thus did "the one for me" become an IBM-PC, the WordStar word processing program, and an Epson printer. Together with my wife, also a professor, we entered the computer age not only having saved considerable money thanks to the university's subsidy but also the agonies of technology shopping.[26]

Regardless of how one came to decide upon a particular computer, once one got it home one realized the truth in Sandberg-Diment's comment that "the 'personal' in 'personal computing' means that once you have bought a machine, the problems are all yours personally."[27] One of these problems was seemingly trivial but in retrospect suggests how naive I and my wife were about a machine that today rules our lives: where to put the IBM? Because we lived thirty miles from the university, we quickly decided it would live at home rather than in one of our Stanford offices. Like so many others we thought of the computer as a kind of turbocharged typewriter, useful primarily for drafting and editing manuscripts but little else. So we decided to install it on a card table in my study, where I was about to tackle the transformation of my doctoral dissertation into my first book. But in so doing, we assumed our new IBM-PC would be an "IBM-SC," or "shared computer," rather than a personal one; we could not imagine that *both of us* would ever need such a machine. The plan was that when I was not working on my manuscript, my wife and I would trade places; she would use the computer in my study while I worked at her desk or elsewhere in the house. It did not take long, however, before I was regularly using the IBM-PC to write lectures, prepare student handouts, take notes from books, and for all the other writing tasks I had once tackled with a pen or typewriter. As the IBM became "personal" for me, I increasingly fought over it with my wife, who thought it was her turn to use the machine as her attitude toward the IBM had simultaneously evolved from minimal interest to increased dependence on it for all kinds of writing. After a year of bickering, we purchased a second IBM; finally, each of us had a truly "personal" computer rather than a problematic shared computer.[28]

Problems of sharing a computer or where to put it paled in the face of the greater challenges confronting early adopters. Just as consumers in the early 1900s brought little relevant knowledge or skill to their first encounters with automobiles, early computer users had little prior experience of value. They may have known touch typing, bookkeeping, and how to work with spreadsheets—they might even have been math whizzes, adept at equations, and natural computer programmers—but new users found the computer itself an alien and strange new monster in the home. Never before had people had to "communicate" with a machine, typing coded commands and hitting "function keys" on a keyboard. Whether people used a computer to write a memo, update a ledger, or do anything else, the result of their efforts existed only ephemerally on the machine's screen and in its electronic innards—unless users "saved" it to a tape drive or floppy disk, themselves alien new devices, or learned to print it out using another new consumer machine, a computer printer, which required entering additional coded commands. In short, using computers demanded a battery of new skills and routines that most people did not find logical or rational. Just remembering the function keys and codes required for using those early machines and their many, seemingly random commands was difficult, and I recall pasting sticky notes with commands written on them all over my machine and adjacent desk. Whether one knew the commands or not, computer routines had to be performed in a specific order and with absolute precision; otherwise the machine refused to do one's bidding. Furthermore, if the machine didn't do what you expected, diagnosing the reason was often impossible, creating another kind of nightmare.

To begin work with a personal computer one obviously had to turn it on, just like any other electrical appliance. Users had no trouble with this—if they could find the switch, which, as on the IBM-PC, was often purposely located out of the way so users would not inadvertently turn it off and lose data. The big challenge came next: dealing with the machine's operating system using the keyboard. In the early 1980s, computers had no mouse, no GUI, or graphical user interface, and no pull-down menus. Instead, one typed instructions at the so-called "command line" that appeared on the black-and-white (or with IBM machines, green or orange on black) screens. The most common operating system was Microsoft's DOS (disk operating system), used by the IBM-PC and many other personal computers. IBM charitably described DOS as the user's "silent partner" in running the computer, yet this partner was difficult and prickly in that its commands required typing codelike strings of alphanumeric symbols. "File operations can be quite overwhelming," wrote one reviewer of DOS, who explained that, given "the long file names sometimes necessary, copying a file from one directory to another may require typing a command like: COPY C:\WORK\MEMO\STATUS.JUN A:\BACKUP."[29] Translated, that command instructed the computer to copy a memo on the subject of the status of June work from one floppy inserted in the C: drive (one of the machine's two floppy drives), onto a second floppy labeled "backup," which is in the machine's A: drive. The backslash key required by DOS commands, along with the Esc, Alt, and Ctrl keys users encountered on most PC keyboards, were unknown on typewriters further adding to the novice's bewilderment.

With its arcane symbols and coded commands, DOS demanded total accuracy and precision. Type a single character incorrectly, be it a period, comma, number, slash, or letter, and the computer would not do what you wanted it to do. Never had a technology been so fussy. The first generation of automobiles permitted users to be somewhat sloppy, starting their engines without retarding the spark, say, or moving out from a dead stop in second or even third gear; the vehicle might react noisily and roughly, but it usually responded. Other seemingly precise technologies, such as the combination locks used on lockers or luggage, also cut users some slack. If the first number of a lock's combination were 32, say, one might dial "31" or "33" and the lock would still open. Personal computers, however, forgave no such imprecision or errors of "syntax," as computer programmers call it. If "33" is part of a command, then users had better type that number or the machine will not obey. A typo might prompt a scary DOS response on the screen, "Abort, Retry, Ignore?" To move ahead users then had to choose one of the options, typing "A," "R," or "I," often not having known what went wrong.[30]

Computer owners found their most used application, word processing, equally tyrannical in its demand for precision. Early word processing programs did not care about content; users could misspell words just as they had with typewriters (automatic spell checking as one types came later); rather, it was in telling the computer to process words and instructing the word processing program what to do with ones writing that required exactitude. To use WordStar on an IBM-PC, for example, whether to write a letter, compose a lecture, or work on any other text, meant first instructing DOS to load the program by inserting a floppy containing the program into one of the machine's drives and then entering a command naming the drive followed by "WS," like A:WS. The disk drive would whirr for half a minute or so and then the blank WordStar screen would appear. If one wanted to work on an already existing WordStar file, the start-up process was a little more complicated; one would instruct DOS to call up and read the earlier writing from a floppy disk the user has inserted into the machine. Once WordStar was loaded and the WordStar screen appeared—black with white letters (sometimes green or amber, depending on the model computer) along with the flashing little horizontal line called the cursor—writing became like typing but with the differences we now all cherish. The faster one typed, the more rapidly words marched silently from left to right across one's screen, wrapping magically at line's end to begin a new one as the text scrolled downward on a seemingly infinite page.

The real magic compared to writing on a typewriter, however—the benefit those of us who adopted computers and word processing celebrated—was the ability to delete, move, and rearrange our words, sentences, or paragraphs. As writers and typists accustomed to editing on paper, we brought to computers a familiarity with the traditional cut-and-paste method of revising that employed scissors and glue or tape. The ability to cut and paste digitally rendered the manual skills irrelevant, though editing with a computer required learning another whole repertoire of commands. These often required striking two or even three keys on the computer keyboard simultaneously, a strange departure from typewriter practice, where typists struck two keys at once only if they wanted to capitalize a letter or, much less frequently, to type a number or symbol (depressing the shift key while simultaneously striking a letter or symbol). Hitting any two other combinations usually jammed the typewriter's mechanism. Computers on the other hand could be programmed so any number of keys simultaneously depressed signaled a different command (only the human hand limits the number of keys that can be hit at once). With WordStar, many commands required holding down two and sometimes three keys at once. The Ctrl key in combination with another key performed all kinds of different operations. Hitting Ctrl+R or Ctrl+D, for instance, moved the cursor respectively one character to the right or one line down, while Ctrl+F or Ctrl+A moved it one word to the right or a word to the left. Those who brought touch-typing skills to computing often preferred the two-key WordStar commands over using the computer keyboard's directional arrows for moving the cursor or function keys for other operations because those commands did not require lifting one's hands from the typist's "home" position, with one's index fingers resting respectively on the F and J keys.[31] Because computer keyboards have a repeating feature—holding down a key or keys causes the machine rapidly to produce the commanded action again and again—typists could use Ctrl+U or Ctrl+D to move their cursors rapidly up or down their screens rather than the directional arrows, or in lieu of the more complex procedure for deleting blocks of text, hold down Ctrl+T, which would delete words one by one to the right of the cursor, rapidly deleting sentences or longer passages of unwanted text.

Without a mouse to highlight the text and then drag and drop it where one wished, so called "block operations"—moving words, sentences, or paragraphs from one place in a manuscript to another—required a separate command for each of them. Users first moved their cursor to the beginning of the block of text they wanted to move and then entered the three-key command Ctrl+K+B Second, they positioned their cursor at the end of the block and entered Ctrl+K+K, which highlighted whatever text was between the two points. Finally, users moved their cursor to the so-called "insertion point" where they wanted to move the text and typed Ctrl+K+V. One might save a few keystrokes when performing these editing tasks by using three of the keyboard's ten function keys WordStar dedicated to block operations, but as noted the touch typist was less likely to use these, as in order to reach them required lifting one's fingers from the home position. Whatever the commands used to move text, we early adopters thought it wondrous to see our words jump instantaneously from one location to another.

Even when our cutting and pasting was done and our writing "perfect," the computer demanded more of us. Unlike writing on a typewriter, or with pen or pencil for that matter, we had nothing on paper when we were finished. If we wanted a "hard copy," as we came to call the work we did on computers when printed on paper, we had to learn to use another machine, and master yet another set of commands and procedures. Word processing programs in the early 1980s used so-called "print codes" to tell printers when to italicize, underline, or produce other special effects such as boldface or superscripts. To enter these codes when using Word-Star, for instance, one again confronted three-key combinations (or used function keys). Typing Ctrl+P+Y told the printer to start italicizing while typing the same code again caused the printer to stop. If one wanted the printout to italicize the term *WordStar,* after entering the codes one saw ^PYWordStar^PY on one's screen.[32] These print codes, like other commands, were abstract, seemingly random, and hard to remember, like a foreign language. But the biggest problem with them was that it was so easy while writing to inadvertently omit one, especially in a longer document with many italicized or underlined items. If this happened, the printer would begin italicizing (or underlining) when it encountered the first code, stop when it found the next one, and then start again on seeing a third and so on, and if one were missing the entire document would be italicized everywhere it shouldn't be. Not surprisingly the recycling bins around my university's academic departments were filled with discarded pages, ruined by our failure to get these print codes precisely correct, causing dumb (but precise) printers to run amok, underlining or italicizing page after page.

THE PRINT CODES EARLY ADOPTERS HAD TO TYPE at the command line on their computer's screens, along with the operating system's own commands, were a wholly new foreign language, difficult to learn and hard to remember. But the codes and commands seemed random and often had no discernable relationship to the concrete tasks they caused the computer to do; they were symbols without referents. This aspect of using early computers had no parallel with the work early automobilists had to do with their new machines. While the language of "Motor" included much strange vocabulary like *carburetor, spark plug,* or *vibrators,* those words all stood for tangible things, components that operators could see and touch, making them easier to learn and remember. The computer is a symbol-manipulating machine, which was both the secret of its tremendous versatility but also the reason, when users had to themselves communicate with it using coded symbolic language, it was so difficult to use.

For computer users in the late 1970s and early 1980s, the floppy disk—or floppy or diskette—provided the closest parallel to the early motoring experience. A thin 5¼-inch-diameter plastic disk encased in a protective cardboard sleeve, the floppy was the common means of storing all computer data, as machines did not yet have hard drives. As a consequence, every time users turned on their computers, they had to insert a floppy containing DOS, WordStar, or any file they had previously created with their machine and wanted to modify, into the computer's drive and load it anew into its internal memory. Before turning the machine off, they copied whatever work they had done from that internal memory back onto another floppy to preserve or "save" it.

If tires were the Achilles' heel of early motor cars, floppies were the equivalent for early personal computing. The disks were easily damaged and their information rendered unreadable by many things in the everyday environment where computers were used. If a floppy were bent or scratched after being placed in one's pocket or briefcase, data could be lost. Floppies also were destroyed by heat, electromagnetic fields, and liquids. Computers themselves generated considerable heat, and users who piled diskettes atop their machines might find that their data was gone. Spilling coffee, soda, or other drinks on disks also could ruin them. Telephones, printers, and electric pencil sharpeners on one's desk emitted electromagnet fields that were strong enough to erase or damage data on nearby floppies, as could static electricity. Just touching the computer after walking on the carpet across the room in rubber-soled shoes might zap a floppy and destroy its data. "Electrostatic Discharge—It Kills, It Maims, It Corrupts," screamed a writer in an Apple user-group newsletter.[33] "Once again," warned another, it is time "to prepare for special disk-handling procedures! Lower humidity levels means that static can quickly become a problem, and you can even more quickly destroy that favorite program if you're not careful."[34] Floppy disks also could be injured by dust and airborne pollution, as could early computers generally. "My disks die at an even greater rate," one owner estimated, "I would guess about 1 out of 10 per year, if not more." The reason? "I smoke."[35] Finally, information could simply disappear from floppies, oozing mysteriously into the ether for no discernable reason.

The vulnerability of fragile floppies prompted many early computer users to take extraordinary precautions. When my wife and I first got our computer in 1982, the tech support staff at Stanford advised us never to insert the original DOS and WordStar diskettes into our machines. Instead we were to make "working copies" and then store the originals somewhere safe; we always could make new copies but replacing the original would be costly.[36] Like most early users, we were very anxious about losing data, so we also backed up everything we did on our computers onto floppy disks. I was so paranoid about losing pieces of my first book manuscript that when I went to the bathroom or kitchen for a cup of coffee, I backed up whatever was in my computer's RAM onto a floppy disk, just to be safe. I also briefly kept *three* copies of each chapter, each on its own, separate floppy: a working copy; a backup that I updated after each work session; and finally an archive copy, or a backup of the backup, which I took to my office on my next trip to the university, thirty-five miles away. I'd heard stories of people losing everything and wanted to make sure that didn't happen to me! Yet the logistics of my regimen were overly complicated and I quickly reverted to keeping a single backup copy of each chapter. A friend of mine so worried about losing her data that she not only backed up each day's work onto floppies but also printed it out, squirreling away reams of hard copy in a file cabinet.

THESE EFFORTS TO PROTECT DATA may seem extreme, but computers had brought a wholly new evanescence to the written word. Unlike texts produced with pencil, pen, or typewriter (or artworks and anything else created by hand using earlier technologies), when the pen's ink ran out, the pencil's point broke, or the typewriter jammed, what already was done was still there on paper no matter the problem. With a computer, on the other hand, what one wrote existed only as spinning electrons visible as print on a screen or, only a little more permanently, as magnetic inscriptions on a thin plastic disk—unless one took measures to produce a paper document using a printer.

The earliest adopters of computers seldom owned a printer, which was the "expensive missing link," as pioneer computer journalist and retailer, Stan Veit, put it, in the vision of home computing. Some hobbyist types built their own printer from a kit or cobbled one together from an old electric typewriter or teletype machine; others, if they printed at all, accessed printers at their workplace, in laboratories, universities, or businesses.[37] By the early 1980s, however, one could buy a printer, such as the popular Epson dot matrix unit we got with our IBM-PC, for a few hundred dollars. But none of us were trained to *print* things—it was a whole new task for consumers. We also learned that dot matrix machines like the Epson were both slow and noisy, emitting banshee-like screams when printing. Also, unlike a typewriter, whose product, the typed page, sufficed for most purposes, the results of dot matrix printing technology did not. This was because each printed character was made up of many tiny dots rather than of a single fully inked letter or symbol that was produced with a typewriter or printing press. And dot matrix paper was also special. It had an extra strip running down each side with sprocket holes (used by the printer to "feed" the paper in and out), while at top and bottom each sheet was attached, accordion-like, to the next; after printing something, one tore the sheets apart at pre-perforated locations and ripped off the side pieces with sprocket holes as well. The end result was a sheet with fine, serrated edges and lacking what we came to call letter-quality printing. Documents of this kind were considered improper for use in business or professional correspondence, grant applications, manuscript submissions, or for other official purposes. In those days, to print letter-quality documents, we copied them onto a floppy, carried the disk to our university, where most departments already had costly laser printers, and printed there.

The early home printer not only produced poor quality results but, as one expert observed, because of the complexities it introduced into a system, a printer was "the vilest piece of equipment that can be connected to a computer." Because a printer was "more-or-less-intelligent," having a built-in computer chip, its own "memory," and software, connecting it to a computer exposed users to all kinds of possible problems.

"While great advances have been made in taming these noisy cantankerous beasts during the last few years," he continued, it is quite clear that "printers continue to plague many of you—programmers and non-programmers alike." The major reason was that " 'standards' for printers are weak or nonexistent," he explained. "The wide variety of equipment and software in use makes it impossible to give specific 'do this' answers to most printer-related questions. The right answer depends on the exact eccentricities of the equipment, software, and connections involved. The bad news," he added, "is that the only person who is in a position to solve YOUR printer problems is YOU."[38]

Whether the printer or the computer, so many of these computer-related problems were just impossible for consumers themselves to diagnose or solve. Hardware and software makers quickly realized this, and as early as 1980 an Apple Computer executive observed that "one of the areas that will have the greatest impact in separating the men from the boys" in selling computers is "the quality of after-sale support" a firm can deliver.[39] That the computer industry invariably spoke of "support" rather than "service" underscored the supine position consumers had vis-à-vis the new technology; they had to be somehow propped up if they ever were to successfully use the new machines. In spite of the difficulties people had with computers and related equipment, manufacturers misleadingly marketed these products as user-friendly, a phrase that as Sandberg-Diment diplomatically put it in the *New York Times,* was a "much abused appellation."[40] Then and now the industry struggles in its efforts to adequately (and inexpensively) support computer users and help them cope with the almost infinite number of problems that that digital technology can have, although computers themselves are better behaved than they once were.

At the outset computer and software sellers followed the pattern pioneered by makers of sewing machines and automobiles: they provided buyers with a "silent instructor," or printed instructions. The "documentation" accompanying early personal computers, as owner's manuals were often referred to in the industry, was "notoriously unhelpful," reports technology scholar Ellen Rose. Manuals tended to be "last-minute productions dashed off by programmers."[41] The manual for the first Apple computer in 1977, recalled one owner, "consisted of thirty photocopied pages, including some handwritten notes from Woz," Steve Wozniak, the programming genius who with Steve Jobs had founded the company.[42] Five years later, in 1982, Sandberg-Diment wrote that "large number" of personal computer manuals were still "as incomprehensible to the neophyte as the entrails of the machines themselves."[43] This situation opened the door for independently published manuals and guidebooks as bewildered consumers, just as they had with the early automobile, sought more comprehensible published advice and instruction on dealing with their computers. How-to-do-it volumes such as *WordStar Made Easy, WordStar in Three Days: What to Do When Things Go Wrong,* and *WordStar Simplified for the IBM Personal Computer* contained information and often included fuller, simpler explanations than the official factory-published manuals.[44] Some would insist, thirty years after Sandberg-Diment's comment, that printed computer instructions are still largely incomprehensible, that computers and their software programs are not at all like early automobiles or other personal technologies that could be understood using silent instructors. People need personal support with computer problems; they need to talk to another human being who can walk them through their difficulties and solve them.

Early on the computer industry attempted to deliver person-to-person support, experimenting with different approaches. By 1980, Apple had set up a telephone hotline to "offer technical assistance to the dealer," who then, the company assumed, would help retail customers solve their problems.[45] Things did not work out as planned; computer buyers got the hotline number and called Apple directly. "Of the 94%" of calls that came from customers to the hot line in April of 1980, complained Apple's in-house magazine, "only 12% are on a subject which the dealer would not normally be expected to know." Only those calls should go to Apple directly, it reminded its dealer-readers, while "the remaining 80% of the calls should be handled by you! THESE ARE ALL QUESTIONS THAT CAN BE ANSWERED WITH INFORMATION PUBLISHED IN THE Apple manuals or in the Application notes."[46] It seemed Apple dealers were no more inclined to read (or able to understand) the documentation they received than were their customers, so consumers continued to overuse the hotline. "In the excitement of purchasing a personal computer and learning to use it," the company again reminded its dealers, "the new owner may not come away with a clear understanding of what to do if he or she needs help."

But he or she must, management emphasized, utilize the "structure" set up by the company.[47] Whatever the computer industry's plans and hopes, however, consumers would continue to seek out the most convenient and user-friendly support they could find.

Meanwhile, personal computers continued to evolve at a rapid clip, making the Apple IIs, Commodore PETs, TRS-80s, and IBM-PCs of the late 1970s and early 1980s soon seem like relics from the Stone Age. Computers became markedly faster, more powerful, and much smaller; indeed, recent laptop computers are more powerful than the computers that enabled the Apollo 11 crew to fly to the moon and back in 1969. Besides their speed and power, modern personal computers have incorporated many other improvements and innovations that have dramatically altered what we do with them and their importance in our lives. As these are so well known, we need only summarize them briefly.

One innovation first appeared on the new 1983 IBM-XT, a 10 MB hard drive. By today's standards, this hard disk was tiny, but compared to a floppy it stored a tremendous quantity of information. Soon, all personal computers had an internal hard drive along with a drive for a floppy or some other form of movable media such as a CD. With a hard drive, computer users could store their operating systems, word processing programs, and any other programs they commonly used on their machines, rather than have to load them anew each time they booted up. They saved their work to the hard drive before turning off their computers. Hard drives also speeded up computer operations. "Until you use one," wrote Jean Mickelson in a newsletter, "you can't conceive of the difference in speed. In short," she concluded, choosing an automotive metaphor to underline the hard drive's superiority, "floppies seem like a Model T compared to a Ferrari."[48] A second crucial innovation was the graphical user interface, or GUI, first debuted in 1984 on Apple's new Macintosh computer.[49] This employed the now ubiquitous desktop, with little icons representing programs, folders, and files and a menu bar running across the top of the computer screen. Using a new kind of controller, the now universal mouse, trackball, or track pad, people could open files, select items from a menu, highlight text and drag and drop it elsewhere, or toss things into a trash can, all by pointing and clicking. When the next year Microsoft introduced its first Windows operating system, emulating the essentials of the Macintosh GUI, few lamented the demise of the old command-based interface.[50] A mouse and GUI became the standard means whereby people interacted with computers.[51]

Probably the most transformative single innovation in personal computing was the development of browsers, the software that enables computers connected to the Internet to access millions of pages of text and images on the World Wide Web. With the introduction of the first commercial browser, Netscape's Navigator, in 1994, followed a year later by Microsoft's Internet Explorer, people now could send and receive e-mail and use the Web. Now, whatever doubts consumers once had as to the utility of personal computers began to disappear. It took about thirty years from the automobile's first introduction for the majority of American households to have a family car, but after only a quarter century personal computers had the same level of domestic ownership. By 2000 they had become not only desirable and affordable but most people thought them a necessary personal technology. Today, we shop, bank, and pay bills; apply for jobs or college admission; take online courses; share photographs; or download music, videos, and software from cyberspace. We can obtain help or advice on computer problems or almost any issue imaginable.

Unlike automobiles, however, which became user-friendly after thirty or so years of improvement, personal computers have still not reached that point—and it is not clear they ever will. Today's computers are definitely more reliable and easier to use than those of the eighties or nineties, but we still can go through hell with these machines, their remarkable technical progress notwithstanding. Too often some little or big thing doesn't work with a computer. The worst thing is that its hard drive can crash, the computer user's equivalent of Armageddon. This rapidly spinning disk is the only mechanical component in an otherwise all-electronic device, and the constant starting and stopping required of it, as well as the knocks it takes when the computer is being moved around, place it at risk. Statistics say the average hard drive lasts three to five years, and a third of all users have lost all of their data at some point in their computing experience.[52] When this happens prudent users might only be *inconvenienced*, for they would have recently backed up important programs and document files onto removable CDs or an external hard drive and would lose little of value. But most people are more complaisant: nearly 10 percent never back up anything, while only one in four does so frequently.[53] Because

today's hard drives have capacities measured in gigabytes or even tera-
bytes, crashes, when they happen, are often catastrophic as people now
store their lives, so to speak, on their computers. Everything from e-mail
and Web addresses to personal calendars, correspondence, tax and bank-
ing records, family photographs, favorite music, videos—all this and
more often exists only as magnetic traces on the hard disk. Its failure can
be like a death.

For many of us, it is the frequent lesser computer troubles, the glitches,
bugs, incompatibilities, and viruses, that prevent the technology from at-
taining the user-friendliness of an automobile. Speaking personally, it
seems as if no week passes when I or my wife do not have some computer
problem, or experience trouble, even if quickly resolved, with our printer,
scanner, or wireless home network. Suddenly something will not copy or
print, or a Web site will not open, or the machine freezes when we are
trying to perform some routine operation, and that little rainbow-like pin-
wheel or hourglass icon that indicates the machine is busy just spins and
lingers forever but nothing happens. A dialog box may finally pop up to
annoyingly tell us: "You are not connected to the Internet," "The printer
is not responding," or "This disk is not readable." But the dialog boxes
never tell you *why* you have a problem or what to do about it. They expect
you to select the "OK" button, as if the impasse were acceptable. It is not,
of course, but one can't continue without clicking on "OK," so one does.

What choice do we have? All of us who use computers are dependent
on this complex, black-box technology that we don't begin to understand,
and therefore we muddle on. Our repertoire of "fixes" is small. We try
once more to access the Web site, print the document, or copy the disk.
Sometimes this "fixes" the recalcitrant machine, but when trying again
fails to work we deploy our second trick: we quit the troublesome program and relaunch it before attempting to execute the task that previously
failed. If it fails, this time we resort to our final and most emphatic tricks:
we turn off the computer altogether and reboot it, or if our problem is
with the Internet and we have a wireless network, we turn the router off
and then back on, hoping this intervention will dispose the computer to
obey. Mysteriously, this ploy often succeeds, although if we have gotten
things to work again we do not have the slightest idea what went wrong,
why all is now fine, or whether the problem will happen again. To be sure,
most of these troubles are little bumps in the road, so to speak, compared

to earlier computer problems, and hardly as onerous as the accidents that befell earlier motorists. But if our few tricks fail to get our computers running again, we pretty much have exhausted our capacity for self-help and must seek support elsewhere; even the most "tech savvy" often confess to being driven "bonkers" by the technology.[54]

It is most people's almost total ignorance of how computers work that underlies their continuing need for technical support, although generations growing up with computers have a greater facility with them than us older users. Over time, the computer industry has devised clever supporting strategies , including the ever-present pull-down "Help" menus offered by computer programs, the now common FAQ Web sites put up by software providers, and a plethora of other Web-based sites to aid puzzled users. But online support obviously requires that one's Internet access be working, which may be the problem. And computer users still have nothing quite like the neighborhood service stations that once adjusted and repaired automobiles, although a number of commercial enterprises have emerged, at least in more populated areas, to bring personal support closer to computer owners. Internet service providers, or ISPs, such as Comcast have IT departments that users can call for help with their wireless networks. The Geek Squad, an arm of the electronics retailer Best Buy, makes house calls, helping bewildered users with stubborn machines, diagnosing and repairing problems, and setting up wireless networks; Apple Computer stores have installed so-called "Genius Bars," where Macintosh owners can make an appointment and bring in their machines for twenty minutes of free help, whether to diagnose a case of trouble or assist users with software. It is not always easy or convenient to avail oneself of these support arrangements, so it is the rare computer user who has not on occasion suffered too long without help with a balky or inoperative machine. But this is a measure of our servitude to the technology, a price we pay for our embrace of and dependence on computers.

And computerization is spreading, infiltrating other household devices by giving them digital interfaces and enhanced functionality while at the same time changing once user-friendly and easily learned machines into newly challenging technologies. In the epilogue and conclusion that follows, we look briefly at some of the user unfriendly offspring of the personal computer and their implications for technology consumption.

The Technology Treadmill

D IGITIZED AND COMPUTERIZED TECHNOLOGIES have relentlessly moved into every corner of our homes and daily lives. From mobile phones to digital clocks, from fax machines to digital cameras, such devices offer wonderful new capabilities, but at the same time these machines can be user unfriendly in the extreme. Consider the lowly thermostat, a device like the round, gold-colored analog units manufactured by Honeywell. Its design has not changed in over half a century and its operation is simplicity itself.[1] The thermostat has a thermometer that tells one the actual temperature in the room, and a control—the outer shell of the round device—used to adjust the heat. To turn up the heat, one simply turns the shell clockwise, a long established behavior for raising or increasing the output of some machine or device, signaling the furnace to send more heat to the room while simultaneously moving the red indicator pointer along a temperature scale calibrated from 45 to 85 degrees. To lower the temperature one turns the shell in the opposite direction. The thermostat has no other control, not even an on or off switch, for it is wired into the house electric system and is ready at any time to adjust the heat. An analog thermometer like this is so user-friendly we use it without hardly noticing.

By comparison, a digital thermostat is complexity personified, as I first discovered in 2005 when my wife and I completed an addition on our house. In the older part of the home we had the familiar Honeywell analog thermostats, but in the new wing each heating zone had a digital thermostat. Although these digital units communicate with the same furnace, they have many more controls and a more confusing interface; indeed, they epitomize the user unfriendliness computerization makes possible. Along with an LCD screen, each has four buttons: one called "menu," another named "item," and two buttons marked respectively with an upward and downward pointing chevron. The buttons do different things, and the LCD screen shows different information, depending on what "mode" the thermostat is in, making it almost necessary to have the owner's manual by one's side when using the unit. Another complexity emerged when I noticed that our five thermostats (one each for the five radiant heating zones we have in our addition, essentially one per room) did not show the same information. In our first winter with the new technology, I went to raise the temperature in a couple rooms and saw that one thermostat displayed "room temperature" as its default while another indicated "slab temperature." I knew that the temperature of the slab (in our case wooden floor, although the programmer called it "slab") governed the temperature of the room. But I wasn't sure how to make the rooms warmer, which was all I wanted to do. The manual that came with the thermostats was impossibly technical, aimed at installers as much as to users, so I experimented, somewhat randomly pressing buttons until I saw a screen called "Slab, adjust temperature"; this seemed right, and I found that by then pressing the Item button, a number registering the temperature appeared and by using the Up arrow button, I could increase that number. I did this, but even after twenty-four hours, plenty of time for the slow-to-change radiant heating to warm the rooms, they remained cold.

I learned more about the dense complexities of my digital thermostats after the HVAC contractor who installed them and the radiant heating system came and tutored me. Although I had been on the right track adjusting the slab temperature, I had failed to "select" my changes by pressing the Item button after keying in a higher temperature. That essential fact was in the manual, but I'd missed it, but even after I learned how to enter my new and higher temperature, we were still shivering twenty-four hours later. It turned out that I had programmed what the system called a "temporary hold" on the original setting, one that only lasted for *four hours on a single* day. This was the result of the new functionality my digital thermostats provided; the units not only regulated heat but also kept track of time and date, allowing the units to automatically change a room's temperature up to six times every twenty-four hours and to do so differently on different days of the week. I could see how dividing the day into six, four-hour periods might make sense for homes with forced air heating, which rapidly responds to changes in thermostat settings; a household with a forced-air system might program its furnace to warm up the house on weekdays at 6:00 a.m. as family members prepared for school or work, to lower the temperature at 9:00 a.m. after everybody was gone, to then raise it again at 3:00 p.m. when people returned to spend the evening, and finally lower it for the night at midnight. Different schedules on the weekend might save energy.

With radiant heat and its slow response time, however the clock/calendar feature of our new digital thermostats was next to useless. What I should have done, as I learned after yet another visit from the HVAC man and more personal instruction in the working of our units, was to program a "permanent hold" on the initial temperature setting and then set a new temperature. Yet this situation was crazy. What was so sacrosanct, I wondered with my anger rising, about the original temperature setting programmed by the manufacturer or the HVAC installer? Why were my changes, those of the homeowner, called "holds" at all? Why couldn't I or my wife just tell the thermostats what temperature we wanted and have the units provide it—until we didn't want that setting, at which point we would reset them?

I asked the HVAC people to replace our digital units with analog thermostats similar to the ones in the older part of our house. That was impossible, we were told, as the central computer of the radiant heating system could only talk to computerized thermostats. The technician from the HVAC firm could reprogram the central computer of the heating system so as to make each thermostat more closely simulate an analog unit, however, and we told him to go ahead and do that. He removed the confusing four-hour time periods, the day-by-day programming capability, the three different kinds of holds (along with "temporary" and "permanent" holds there was a "vacation" option) and a few other features we would never use in a million years. All our digital thermostats now display the same default information: room temperature rather than slab temperature, and to raise or lower the temperature we first press the "Menu" button to access the "Adjust Room" submenu and then press the up or down buttons to change the setting, and finally press "Item" to select the change we just made (it turned out that all along there were two sensors in each heating zone, one in the slab or floor and the other, essentially a thermometer in the thermostat itself, registering the ambient local temperature). In using our digital thermostats we now pretend there is no slab and just adjust the room temperature, although to do so requires more steps than with the older analog units—and often a flashlight to read the dim, digital information on the screen. By lobotomizing our new digital technology, we destroyed much of its functionality, but we made it more user-friendly.

The same sort of frustrations encountered with those thermostats repeat themselves with other digital devices that consumers of late have brought into their homes and lives: microwave ovens, telephone answering machines, HDTV and home theater equipment, automatic sprinkler systems, digital cameras, air conditioners, bathroom scales, and more. One problem is that the controls on digital interfaces are often confusingly labeled or lack any identification at all. "What does this button do," people wonder, and "how about that one?" If there is a menu or menus, as usually there are on these machines, we often cannot readily understand the options and choices. It isn't that the words are new or difficult—it is rather that a phrase like "permanent hold" may mean little or nothing to consumers who have never encountered it before. Second, the function of the buttons on digital interfaces changes depending on the current state or mode of the device; often a button needs to be pressed in a particular sequence along with other buttons on the interface, and these patterns that can seem extremely arbitrary. Third, even when users know what a button is supposed to do they often are unsure whether pressing it suffices to produce the desired result; some devices emit an affirming beep or provide other positive feedback, such as a message that appears on their LED screen, confirming that things have "worked." Others, like my digital thermostats, do not confirm a user's action and thus create confusion. A fourth and significant source of confusion is the sheer number of options computerized digital technologies make possible, the much lamented "feature-creep," as some have called the phenomenon. Finally, the rapid obsolescence of digital technologies is problematic: people barely learn how to operate a new device before it is rendered obsolete by a new model or upgrade with more features. New versions of older technologies, especially digital upgrades, may well be more user-friendly than the units they replace, but because they invariably incorporate more features they force users to ascend yet another learning curve that can be frustrating

The first common analog device to go digital was the wristwatch. For centuries, watches had been large, pocket-sized devices, but by 1900 they were sufficiently miniaturized that men and women began wearing them on their wrists. A wristwatch's interface consisted of a face with hands coupled with a single control, a stem with a knurled knob or "crown" at its end. To wind a watch one turned the knob, thereby storing energy in the spring within. To change the time or reset the watch, users simply pulled the crown outward, causing gears inside the mechanism to move the hands when the knob was turned. After setting the watch to a new time, users pushed the stem back in to its default, or starting, position. These procedures were as easy and simple as using an analog thermostat, and by the time wristwatches started to go digital in the 1970s, at least three generations of consumers had become comfortable with the traditional interface. Many people learned to manage a watch as teenagers after receiving one as a graduation present or on some other special occasion. Whenever one learned to wind and set a watch, the know-how was transferable to all analog timepieces acquired in the future.

With the advent of digital watches, users faced a computerized interface that was far more complicated and notoriously user-unfriendly. Part of the new complexity was that unlike most analog watches, of which only the most expensive did more than tell time, digital watches commonly tell the day, date, and year and also had stopwatch and alarm functions. Powered by batteries, digital watches do not require winding and, because they use oscillating quartz crystals and computer chips to regulate the time, they seldom need to be corrected. But to set such a watch can be a nightmare. The user must push two or more buttons in precise sequences while reading a small LED or LCD screen, rather than interacting with the familiar stem, hands, and face of an analog watch.[2] The instructions that came with an early digital watch now in the Smithsonian, a 1980 lady's model, were typical. Under the heading "setting your watch," users were instructed as follows: "With a thin, pointed instrument (toothpick, pencil, etc.), push button S2 *two times* to show and freeze the *month* in the display." They were then instructed to push the "S1 button repeatedly" to advance the month and the S2 button once to set the date, which could be adjusted to advance to the correct date by pushing the S1 button repeatedly. A similar process was repeated to set the hour, then the minute and, finally, "to complete the setting sequence, push S2 button once. The display will show and freeze the time you have set. Using another clock or official time source as a reference, push S1 button a final time to activate the watch at the precise moment. The two dots in the center of the display will then begin blinking again, signifying your watch is now set."[3]

Once set, digital watches would keep accurate time, but when their batteries died or when owners needed to adjust to a new time zone or for daylight saving, they had to repeat the sequence of button pushing and most likely had forgotten the procedure and misplaced or discarded their printed instructions. At such points, the interface of a digital watch could become newly baffling.

As with a computer, users have no idea how a digital watch works; it is an inscrutable black box (sometimes literally black), its two buttons— cryptically referred to in the instructions as S1 and S2—too small to be labeled on the artifact itself. Even if they were labeled, their function would remain unclear for the buttons display what engineers call nondeterminism; that is, there is no way one can determine what they do from their appearance or placement.[4] Users are thus at a loss knowing which button to press first and what to do next. Additionally, like keys on a computer keyboard, the buttons do different things depending on the *sequence* in which they are pressed and the state or mode the watch is in at that moment. Because the size of a watch practically limits the number of buttons to two or at most three, these few controls must be able to set and manage the many more functions manufacturers can easily and cheaply build into these computerized timekeepers. As a result the digital watch interface remains a model of user unfriendliness and puzzlement. Even an authority on computer interfaces admitted that he "never could discern any logic to the use of these buttons in controlling the watch's functions."[5] Not a few people who tried digital watches have returned to analog quartz watches, battery-powered and very accurate but having a traditional hands and face plus, most importantly, a simple, old-fashioned, knurled-knob interface.

Consumers who disliked the way a digital watch told time, were repelled by its formidably complex interface, or simply preferred mechanical watches always had a choice, as it never killed off or rendered obsolete its analog predecessor. This has not been the case with many technologies. Some formerly analog devices are now available only with digital controls, while altogether new technologies are often computerized and digital from the start, as was the videocassette recorder, or VCR, which took off simultaneously with personal computers in the 1980s. VCRs were in part like record players, simply electromechanical devices that could play back a prerecorded tape rented from a local video store, showing the movie on one's TV set. Users might have to master the VCR's "tracking control." As the owner's manual for an RCA unit explained this, if "black or white streaks" appeared on your screen, you had to "turn Tracking Control knob slowly in either direction until streaks disappear."[6] But the greater challenge posed by the new technology, essential to its primary purpose, was to program the VCR to *record* a television program. The first step in this programming process was setting the VCR's digital clock. This computerized timekeeper told the machine when to begin and when to stop recording. But setting the clock, like setting a digital watch, involved many steps that had to be performed in precise sequence using just a few buttons. That the challenge was more than many buyers could handle was attested to by the blinking "12:00" on the front panels of VCRs in living rooms everywhere, the intermittently flashing, factory-default setting that alerted owners that their clocks had not been set and that therefore their VCR could not be counted on to automatically record anything.

Programming VCRs so stymied people that there was much joking and discussion of the challenge. Only two-year-olds could set a VCR's clock, one comic quipped, while a *People* magazine article, "Taming the #!*?@!! VCR," captured the level of frustration the technology's digital interface engendered.[7] Engineers soon devised a technological fix for the clock-setting problem, but it only added to the confusion. They developed VCR clocks that were set automatically by timing signals embedded in television broadcasts, but viewers discovered that their clocks would be accurate in the morning, say, but then suddenly in the afternoon or evening mysteriously be one, two, or three hours off. Human factors engineer Kim Vicente explained the cause of this strange behavior: when the Fox TV network "started including a timing signal on the feeds being broadcast from its Los Angeles station," those programs when shown on local stations in the Mountain, Midwest, or Eastern time zones, caused

VCR clocks to automatically reset to Los Angeles time. After the L.A.-derived programs ended, the clocks picked up a timing signal from a local station and correctly reset themselves to local time, leaving owners baffled.[8]

Today, digital clocks are everywhere. While a kitchen fifty years ago might have had a single analog wall clock, today there are likely to be digital clocks on the microwave and regular ovens, the telephone answering machine, radio, and coffeemaker; others are on our cable TV boxes, thermostats, in our automobiles, and most ubiquitously now on our cellphones, where for many users they are supplanting wristwatches. Check into a hotel room, and invariably your alarm clock or clock radio is digital, each with a slightly different interface and procedures for setting the time and alarm, promising to confuse us and guaranteeing that, whenever possible, we will rely on the front desk for a wake-up call or worse be woken up by the alarm that a previous occupant of the room had set for 3:00 a.m. to catch an early plane.

The digital camera may best exemplify the mixed results of computerization for consumers. Its basic technology evolved out of earlier research related to television, but by the late 1980s cameras that sensed light and converted images into computer data files were appearing on the market, about a century after George Eastman introduced the first Kodak snapshot camera in 1888. The earliest digital cameras were pricey, but over the course of the 1990s their cost rapidly fell and by 2000 consumers were abandoning film cameras and purchasing digital replacements. The switch was extraordinarily rapid: film sales peaked in 1999 and within a decade declined to almost nothing.[9] People loved the instantaneousness of digital, of being able to see their photographs immediately and then being able to share them easily with friends by e-mail or on the Web. Using these cameras, however, like any new technology, demanded new knowledge and learning. Even the compact digital cameras with a "Ph.D.," meaning "push here dummy" as wags put it, are hardly for dummies, at least compared to older, mechanical film-based cameras.

The original Kodak snapshot camera, it may be recalled, was sold with the catchy slogan, "You press the button, we do the rest." The slogan only slightly exaggerated the simplicity of the Kodak's interface in that users, after triggering the shutter to take a picture, did have to turn a winder to advance the film into position for the next one. Subsequent film cameras added controls for changing the film speed (ASA); adjusting the aperture, or lens, opening; focusing the lens; and for regulating shutter speed. And many film cameras soon had built-in light meters (which meant dealing with a battery) and a switch to turn the feature on and off. But this was about it. Even more complicated and costly film cameras, such as the Nikon SLR I bought in 1968, had only half a dozen or so buttons, levers, and other controls; one could adjust them in any order, and each governed only a single function of the camera. It was not that difficult to learn to utilize all of the controls of such a camera.

With a digital camera, even the simplest point-and-shoot models, users encounter a more complicated interface. There is always an on-off switch, an LCD screen, and at least six or more buttons. Typically, as on computer keyboards, four of them are directional, enabling users to move around in the camera's menus and scroll back and forth while viewing pictures already taken. A fifth button labeled "OK" usually sits in the middle of the four directional buttons and is used to select options in menus. Additionally, most digital cameras have a zoom lens, which includes another control button. If the digital camera is an SLR, like the Nikon D-5000 I bought in 2009, a direct descendant of my 1968 Nikon, a user confronts no fewer than twenty-three buttons, switches, dials, and other knobs on the camera's body. Learning what each of these does is a formidable challenge, especially since, as with all computerized technologies, many of those controls do more than one thing, depending on the camera's mode or state. The flash button on my first digital camera, a 2003 Olympus, was also the button used to move to the right in a menu or to view a more recently taken picture; the macro button, used for close-ups was also the left directional button. Even if these buttons for moving a cursor up, down, left or right perform multiple functions, their now common arrangement in a quadrangle pattern helps users navigate the camera's interface. On more costly and larger cameras, like my Nikon DSLR, the plenitude of buttons allows more of them to be dedicated to but one task, such as trashing a poor picture or viewing previously taken images, again making learning and operation easier, although the overall complexity of such a camera is far greater.

Just press the menu button on a digital camera and the complexity of the technology reveals itself. Even when menus are well designed and thought out, and their structure clear, the multiplicity of menus and submenus along with the new terminology they introduce and the seemingly infinite options and choices they present to the photographer can be daunting. As with other digital equipment, precision and sequence are imperative elements in working with menus and making the camera work for you. A menu-driven interface has numerous advantages, one being that users can enter information into the device's memory without a keyboard, even alphanumeric data. On initially setting up a camera, one is prompted to input the date and time, a tedious process achieved with the directional arrow keys which move a cursor through the letters and numbers in the menu. Later, one might use the same process to title photographs or name folders in which to store them. Perhaps most importantly, a menu-driven interface allows users to choose among many more options than would be possible if each were controlled by a physical button or device. Usually, and this is part of the problem with small-screen devices like watches or cameras, menus are "layered" so users can never see all the possibilities or options at once.[10] So pressing the "menu" button generally brings up a main menu and users must then select the appropriate submenu to find and choose the feature or option they want. On my digital Nikon, the main menu lists six subordinate (but called "primary" menus); selecting one of them then opens a submenu that itself has more deeply layered submenus offering additional options.

"Menuland," as one camera writer has called the well-designed but complicated menu structure of the Nikon, is a "vast thicket" that users must negotiate if they wish to take advantage of the camera's features. They can change almost everything about how the camera works from the way it focuses to how it "sees" colors, placing what amount to virtual filters on the lens to achieve special effects.[11] They may select from no less than nineteen preprogrammed "Scene Modes" that automatically change the camera's settings for optimum results in photographing everything from "pets" and "food" to "night scenes" and "portraits." After they have taken a picture, users can remove red-eye in a portrait or increase the contrast in a landscape, changes possible with film cameras only in darkrooms and with earlier digital cameras only before shooting the picture. Digital photographers have always been able to rework an image using their computers and software like Photoshop, but increasingly they can do so using their cameras alone. Few may choose to alter pictures this way, but these new capabilities typify the way features are added to any digital technology. They often do not raise its cost in dollars but do add to the cost in time spent learning by users to fully understand and control their device.

Entering "Menuland" of even a simple point-and-shoot digital camera can be frustrating, as I learned soon after acquiring a new Olympus in 2003. Like other digital cameras it had an electronic flash that automatically triggered if the light were insufficient; indeed, that was the camera's default setting. I was visiting a cathedral in Europe where flash photography was prohibited and wanted to change the default setting and turn off the feature. Unlike some cameras of this type, mine had a dedicated "Flash" button that allowed one to access the flash menu and change the flash settings without having to deal with the regular menus; when one pressed the button, a dialog box briefly appeared on the screen revealing the present setting. Pressed the first time the box said, "Auto," the default setting in which the camera itself chose whether to flash or not. To change that setting, however, and this was the nonintuitive and confusing part, I had to press the button four times in rapid succession. This caused the camera to scroll through the four flash menu options: "Auto," "red-eye," "flash on" (when one wanted flash on every shot), and finally, "flash off," the choice I wanted. The camera set itself automatically, without one having to select anything, to whichever setting you last saw on pressing the button. So once I saw the "flash off" menu, I'd selected that option without doing anything more and almost immediately the dialog box or menu disappeared. With my flash supposedly off, I wandered about the cathedral contentedly taking pictures. Then suddenly, I took another one and my camera flashed, causing a number of heads to turn accusingly my way. It turned out that I had only temporarily changed the "flash off" setting (shades of my digital thermostats!). The setting was not "sticky," in digital photographer–speak, meaning that my change did not stick if the camera was turned off or went to sleep, the latter being what happened to me in the cathedral. In such cases the setting reverts to the default, "Auto." Those who designed the Olympus gave users no way to permanently disable its flash.

Life in default mode is not all bad. Most of the time the preprogrammed intelligence of a digital camera, left to its default devices, makes pretty good photographs, better on average than many users would have gotten setting the camera's variables themselves. As the manual for my Olympus point-and-shoot camera puts it, "The camera automatically sets the optimal shooting conditions," self-adjusting for exposure, shutter speed, focus, and even the sensitivity of the electronic sensor that replaces film.[12] We may at times quibble with the definition of "optimal"—the camera's flash behavior in the cathedral was lousy—but most of the time we are content to live with the default settings of our digital devices.

A more troubling "default" characteristic of digital technologies is their short lifespan. It is not that digital cameras or other such devices quickly fail or even wear out; compared to mechanical technologies they are fairly reliable. It is rather that digital devices become dated far more rapidly than earlier machines. The phenomenon of "technological obsolescence," as it is called, occurs when over time a machine is so improved in performance, construction, or ease of use, or it is supplanted by some wholly new device, that it becomes all but useless: technologically obsolete. There are instances where this happened in the nineteenth century, but it was the automobile in the early twentieth that first introduced consumers to continuing, unrelenting technological obsolescence. "Why can't I have a permanent automobile as well as a permanent typewriter?" inquired a man writing to the editor of an automobile magazine in 1915.[13] He had bought a new car just three years earlier but already felt that the performance of the latest models so surpassed that of his vehicle that he was upset. There had been a time a few years earlier when some automakers had retrofitted previously sold vehicles with the latest improvements, even doing so without charge, but those days were over.[14] By 1915, there was no such thing as a "permanent" automobile. But there was a permanent typewriter, or so the correspondent believed, presumably because the design and performance of that technology had not changed in years. Somewhat like clocks and sewing machines, typewriters had initially gone through a period of relatively rapid technical evolution and improvement but then reached a kind of plateau where machines were technically more or less the same year after year. Even when the pace of technical innovation in automobiles slowed somewhat in the late 1930s, it was impossible for consumers to think of a car as permanent.

By that time, a different kind of obsolescence began to characterize many technological products—planned, or stylistic, obsolescence. In the late 1920s and into the 1930s, Detroit institutionalized planned obsolescence by introducing the annual model change.[15] Manufacturers deliberately altered the appearance of their vehicles each year by restyling the grill, hood, headlight surrounds, bumpers, trim, interior appointments, and other cosmetic features of the car while leaving its mechanical and functional components, everything under the sheet-metal skin, relatively unchanged. By introducing a "new" model every year, previously sold vehicles, even those just a year or two old, would seem out of date and less desirable, or so the industry believed. Alfred E. Sloan, the head of General Motors and one of the key architects of planned obsolescence, "did his utmost to find new ways to decrease durability and increase obsolescence," writes historian Giles Slade in his recent book, *Made to Break: Technology and Obsolescence in America*.[16] Slade's claim is true with regard to obsolescence but less persuasive to the effect that the cars of GM or any other manufacturer were, in the words of his provocative title, "made to break." The contrary was true, in fact, for by the latter 1930s, just as technological obsolescence in the automotive field was being supplanted by stylistic obsolescence, cars were lasting longer and breaking less frequently. Vehicles built just before the Second World War may not have been "permanent," but they endured longer than those made before the First World War and never became obsolete with the rapidity computerized, digital technologies would.

The appearance of personal computers in the late 1970s and early 1980s again brought to the fore rampant technological obsolescence. As many readers will remember, during the early years computing equipment became obsolete not just in a few years but sometimes in months. In 1983, comparing the life cycle of computers with that of automobiles, Erik Sandberg-Diment of the *New York Times* addressed the accelerated technological obsolescence of the new machines. "The discrepancy in the aging process between the engines of transport" and the technology of "information handling," he asserted, all "has to do with function." The functionality of a car is enduring, he wrote, taking as an example a popular, postwar British sports car to develop the point. "Not only is a 1949 MG beautiful to behold," he argued,

> but it also remains, for the most part, a practical and enjoyable means of transportation. A 1977 Sol may still be one of the most visually pleasing of all personal computers, but its usefulness has declined so far it's probably more suited to anchoring a canoe than computing.

Obsolescence in computer technology marches apace, and it is probably responsible for more of the questions I receive than any other
facet of computing.[17]

Given the MG's notorious reputation for electrical problems, those knowledgeable about sports cars may quibble with Sandberg-Diment's quali-
fier, "for the most part." But his basic point was valid—that a nearly
thirty-year-old car, especially if it was restored, was still a "practical and
enjoyable means of transportation"— while a six-year-old (or even a six-
month-old) computer in 1983, no matter its condition, would have been
all but worthless, technologically obsolete and useful perhaps only as an
"anchor" or doorstop.

In 1983, one might have used a 1977 computer for the same purposes
it was originally purchased: to play games, program in BASIC, compose
short memos or letters, and a few other things. But consumers expected
to be able to run spreadsheet programs like VisiCalc, word processing
packages like WordStar, and personal finance software such as Quicken
on their computers, none of which the old 1977 machine could handle
even if upgraded. More than likely its owner had already consigned the
obsolete machine to the clutter of his garage or sent it to the local land-
fill. The old computer was a victim of the ruthless dynamic that, though
having precedents earlier in the twentieth century, only became endemic
with the digital revolution. By the mid-1990s, as computer users increasingly went online, the pace of technological obsolescence only increased.
Since then, computers and software increasingly have had to be compat-
ible with the demands of other machines and servers on the Internet, as
people expect to send and receive e-mail messages, upload and download photographs and videos, interact on social networking sites, shop
and bank online, and protect themselves from the legions of criminals
and scammers inhabiting cyberspace. Such expectations meant regularly
upgrading both computers and software, a process made more urgent by
the habit of synching smartphones with or attaching digital devices like
cameras, MP-3 players, and scanners to their computers.

To date, the personal technology that becomes obsolete the most rapidly is the mobile phone. On average, Americans replace their phones
every eighteen months, more frequently than either cars or computers.[18]
Even my wife, a relative technophobe and late adopter regarding many

technologies, fits this pattern. She got her first cell phone in 2002, up-graded to a new one in 2005 and then again in early 2008. Then in the autumn of 2009, struck by how many of her professional friends were using iPhones, she upgraded once again, acquiring her fourth mobile in seven years. While some replace their phones for largely fashion reasons, my wife was motivated primarily by the iPhone's ease of use and capability of sending and receiving e-mail, essential to her work and something she could not do with her previous phones. Her previous mobile still worked, but for her it was technologically obsolete, being incapable of doing what the newest generation of phones routinely provided.

Thirty years ago, Sandberg-Diment warned about the rapid obsolescence of digital technology. "A computer is not something you'll acquire once and keep forever," he wrote, and "once you're on the technology treadmill, there's no getting off until you drop."[19] His prediction proved prescient, for the treadmill is more crowded than ever. There is virtually "no group out of the tech loop" in American society today, asserted a 2009 study reported in the *New York Times*. Young and old, male and female, "we're all gadget geeks now."[20] Most of us own (or desire) mobile phones, personal computers, and HD TV sets; moreover, we have (or covet) gadgets like Wiis and Xboxes, digital weather stations, computerized sewing machines or programmable woodcutting machines for our personal hobbies and avocations. While these technologies are usually acquired according to the demographics of age and gender, they ensure by their diversity that virtually all of us are technology consumers, motivated to improve or modernize our lives and lured by the continuing seduction of gadgets that promise to entertain us, make us more productive, save energy, or ease some aspect of our labor.

That we struggle with this ever-expanding technological world is attested to by all the humor that makes light of our plight. *New Yorker* cartoons about computers—an example of which was noted in an earlier chapter—frequently report on how we are tethered to the technology treadmill. In a 1991 example a woman who has come into a bank and holds a shotgun to the teller's window says, "Stand aside, Gruenwald! It's the computer I'm blowing away!" Another cartoon from 1996 shows a woman, waiting for an elevator, pondering its control panel's three buttons: "Up," "Down," and "Other Options." In yet another example from 1997, a man stands in front of his microwave: "No, I don't want to play chess. I just want you to reheat the lasagna," he says.[21] We chuckle because such cartoons mirror, with only the slightest exaggeration, the frustrations we all have experienced in our relationships with our personal technologies. We recognize the blind rage that might prompt one to destroy a machine, to blow it away; we too have puzzled over options that make no sense whatsoever; and everyone of us has a story about just wanting "to reheat the lasagna" while being foiled by some technology that in effect thinks it knows what we want better than we do. The variety of user unfriendly behavior our machines are capable of is infinite. Yet we are dependent on these devices, be they wholly digital or old-fashioned mechanical, and cannot live without them. Even if they offer us far more options than we need and are user unfriendly, we have little choice; we stay on the technology treadmill.

acknowledgments

I owe thanks to a generation of archivists, librarians, and curators who have helped me along the way. I began my research at the Ford Motor Company archives at the Henry Ford Museum (now known as The Henry Ford), in Dearborn, Michigan, where I was fortunate to meet Dave Crippen, then the head of the archives, who shared his encyclopedic knowledge of the collections and turned me loose in the acres of stacks to conduct my research. I returned many times to the Ford archives, and for nearly two decades have been kindly assisted by other staff there, including Steve Hamp, Linda Skolarus, Patricia Orr, Cynthia Reid, and automotive curators Randy Mason and Bob Casey. All have helped with this book.

As my research broadened beyond automobiles to embrace nineteenth-century technologies such as clocks and sewing machines as well as late-twentieth-century personal computers and other devices, I incurred debts to staff at other libraries and museums. I want to thank Kathleen A. Zar of the John Crerar Library at the University of Chicago; Kim M. Miller of the Antique Automobile Club of America Library and Research Center, in Hershey, Pennsylvania; John B. Straw at the Library at Ball State University; Stuart McDougall, who presided over the automobile research collection at the Free Library of Philadelphia; Henry Lowood and Maggie Kimball of Stanford University's Special Collections and Archives departments; Chris Cotrell, Jim Roan, and Stephanie Thomas of the library at the Smithsonian's National Museum of American History (NMAH).

Of particular importance in my work has been the NMAH. It has over many years provided me with intellectual stimulation and financial support during various sojourns there as a researcher, most recently during the 2006–7 academic year when I was a Fellow at its Lemelson Center for the Study of Invention and Innovation and then a Senior Smithsonian Fellow at the museum. My project adviser at the NMAH, Roger White, generously shared materials from the Transportation Division's collections and made possible a visit to the museum's automobile storage facility in Maryland. Getting behind the wheel of those early vehicles, even though not operative and in a warehouse, allowed me to better imagine and write about the challenges drivers faced eighty or more years earlier. Carlene Stephens similarly enabled me to study firsthand early clocks and watches in the museum's collection, just as Hal Wallace and David Allison did with VCRs and early computers, respectively; Heather Paisley-Jones of the Education Department loaned me an old foot-powered sewing machine so I could learn for myself how tricky working a treadle is while also feeding fabric under the needle; sewing machine curator Barbara Janssen, aided by seamstress-curators Maggie Dennis, Alison Oswald, Kay Peterson, and history intern Amy Isaacs, participated in a "sewing bee" that I organized in which we all examined a group of early-nineteenth-century sewing machines in the museum's collections, an exercise that answered many questions about that early household technology.

Over an extended period I have also had profitable conversations with many others at NMAH and want to thank in particular John Fleckner, David Haberstich, Craig Orr, and Alison L. Oswald of the museum's Archive Center staff; Joyce Bedi and Art Molella of the Lemelson Center; and curators David Allison, William L. Bird, Michelle Delaney, Paul Forman, Bart Hacker, Steve Lubar, David D. Miller, Jeffrey K. Stine, Carlene Stephens, Gary Sturm, Margaret Vining, Bill Withuhn, Helena Wright, and the late David Shayt.

Numerous other colleagues and friends have also given me valuable suggestions, leads to materials, or just moral support, and I want to thank those I can remember while apologizing to those whose names I may have forgotten: Aaron Alcorn, Reggie Blaszczyk, Kevin Borg, Lonnie Bunch, Jonathan Coopersmith, Rolf Diamant, Susan J. Douglas, Kathleen Franz, Brian Horrigan, Peter L. Jakab, Elizabeth Johns, David Kirsch, Bob Loughlin, David N. Lucsko, Stephen L. McIntrye, Clay McShane, Alan Meyer, Nora Mitchell, Arwen Mohun, Erik Rau, Michael Schiffer, Bruce Sinclair, Rebecca Slayton, Roe Smith, Mark and Stephanie Weiner, and Roz Williams. Even though we had and have never met, Roger C. Allison, a collector of Winton automobiles, corresponded with me at length about the company's early vehicles and generously shared copies of technical publications and other materials.

Other friends and colleagues have read all or part of the manuscript and contributed to making it a better book, so I want to express special appreciation to Paul Ceruzzi, Robert McGinn, Bob Post, Rudi Volti, and Bill Youngs. Family members have also offered helpful suggestions on or read portions of the work, including my brother, John H. Corn; my brother-in-law, Keith M. Jones Jr.; and a distant cousin, John Goldsmith, a longtime employee of the technical handbook and trade journal publisher McGraw Hill, who also put me in touch with the company's archivist, Mary Pearson.

I also want to acknowledge a National Endowment of the Humanities Travel to Collections Grant that, early on in my research enabled me to survey the vast terrain of the Ford Motor Company collections in Dearborn, while a yearlong fellowship from the Stanford Humanities Center provided time for unfettered thinking at a point when the project was beginning to jell as a book.

Finally, I want to acknowledge the tremendous debt I have to my wife. Her skills and sensibility are evident on virtually every page of the book. While its deficiencies are mine alone, whatever strengths it has owe much to her tireless efforts and constant moral support: "You can do it!" Without her the book would not exist, so it is with unbounded love, admiration, and appreciation that I dedicate *User Unfriendly* to my best friend and reader-in-chief, Wanda M. Corn.

notes

Abbreviations

AACA	Antique Automobile Club of America Archives, Hershey, Pennsylvania
ACR	Apple Computer Records, Stanford University
BL	Bentley Historical Library, University of Michigan
CRH	Charles Roswell Henry Papers, Bentley Historical Library, University of Michigan
CSAA	California State Automobile Association
DPL	Detroit Public Library
DV	Sam De Vincent Collection of Illustrated American Sheet Music, National Museum of American History, Smithsonian
FA	Ford Archives
HBL	Baker Library, Harvard Business School
HL	Hagley Library, Wilmington, Delaware
JCL	John Crerar Library, University of Chicago
NMAH	National Museum of American History, Smithsonian
NYPL	New York Public Library
PCA	Personal collection of author
SA	Studebaker Archives, South Bend, Indiana
SMCR	Singer Manufacturing Company Records, WSHS
WC	Warshaw Collection of Business Americana, NMAH
WSHS	Wisconsin State Historical Society

Introduction: Our Marvelous and Maddening Machines

1. Edward Tenner, *Why Things Bite Back: Technology and the Revenge of Unintended Consequences* (New York: Knopf, 1996), 7, and generally chap. 1, "Ever Since Frankenstein."

2. Gary Cross, *An All-Consuming Century: Why Commercialism Won in Modern America* (New York: Columbia University Press, 2000), exemplifies this tendency, though it is one of the best studies of twentieth-century consumption. Other sig-

nificant studies of consumerism are William Leach, *Land of Desire: Merchants, Power, and the Rise of a New American Culture* (New York: Pantheon, 1993); Susan Strasser, *Satisfaction Guaranteed: The Making of the American Mass Market* (New York: Pantheon, 1989); and the essays in Simon J. Bronner, ed., *Consuming Visions: Accumulation and Display of Goods in America, 1880–1920* (New York: W. W. Norton, 1989). A study that is very sensitive to the subtle distinctions of different kinds of material culture, though it deals with a period before the industrial era that concerns us here, is Richard Bushman, *The Refinement of America: Persons, Houses, Cities* (New York: Random House, 1992). An important study focused on technology, although interested primarily in consumer choice, is Ruth Schwartz Cowan, "The Consumption Junction: A Proposal for Research Strategies in the Sociology of Technology," in *The Social Construction of Technological Systems*, ed. Wiebe E. Bijker, Thomas P. Hughes, and Trevor Pinch (Cambridge, MA: MIT Press, 1989), 261–80.

3. For examples of studies sensitive to consumption practices beyond shopping and purchasing, see Lizabeth Cohen, *Making a New Deal: Industrial Workers in Chicago, 1919–1939* (Cambridge: Cambridge University Press, 1990); Ronald R. Kline, *Consumers in the Country: Technology and Social Change in Rural America* (Baltimore: Johns Hopkins University Press, 2000); and Ted Ownby, *American Dreams in Mississippi: Consumers, Poverty, and Culture, 1830–1998* (Chapel Hill: University of North Carolina Press, 1999).

4. Susan Strasser, "Making Consumption Conspicuous: Transgressive Topics Go Mainstream," *Technology & Culture* 43, no. 4 (Oct. 2002), 755–70, 762.

5. David M. Potter, "American Women and the American Character," *Stetson University Bulletin* 62 (1962), 14.

6. Among historians of technology, studying "users" has recently become popular. See Nellie Oudshoom and Trevor Pinch, eds., *How Users Matter: The Co-Construction of Users and Technologies* (Cambridge, MA.: MIT Press, 2003); Ron Kline and Trevor Pinch, "Users as Agents of Technological Change: The Social Construction of the Automobile in the Rural United States," *Technology & Culture* 37, no. 4 (Oct. 1996), 763–95; and Steve Woolgar, "Configuring the User: The Case of Usability Trials," in *A Sociology of Monsters: Essays on Power, Technology, and Domination*, ed. John Law (London: Routledge, 1991), 57–99.

7. Sharon Zukin, *Point of Purchase: How Shopping Changed American Culture* (New York: Routledge, 2004), 169.

8. Ibid., 29.

9. Ibid., chap. 4, 89–112.

10. Historians of technology will detect their scourge of "technological determinism" in my argument here, yet I believe a machine's interface and intrinsic needs severely restrict the freedom a human being has as a user; to this extent,

then, machines determine *how* they must be used if not the *purposes* for which they are used. Put differently, what owner/users *do* with their machines, and whether or not they alter them, is not technologically determined. For an overview of this debate over technological determinism, see Merrit Roe Smith and Leo Marx, *Does Technology Drive History: The Dilemma of Technological Determinism* (Cambridge, MA: MIT Press, 1994).

11. Abbott Payson Usher, *A History of Mechanical Inventions* (New York: Dover Publications, rev. ed., 1954), 116.

12. On the subject generally, see Judith A. McGaw, ed., *Early American Technology: Making and Doing Things from the Colonial Era to 1850* (Chapel Hill: University of North Carolina Press, 1994).

13. Michael A. Bellesiles, *Arming America: The Origins of a National Gun Culture* (New York: Knopf, 2000), 18. Bellesiles's work not only inflamed the already ongoing political struggle over a citizen's right to bear arms under the Second Amendment but came under criticism from scholars and reviewers who claimed his use of evidence in his book was at worst misleading or fraudulent and at best sloppy, claims that eventually caused him to lose his professorship at Emory University. Bellesiles's book and practices spawned a considerable body of critical and polemical literature, not least among legal scholars. See, for example, John J. Donohue, "Guns, Crime, and the Impact of State Right-to-Carry Laws," *Fordham Law Review* 71 (2004), 623–52. In spite of the controversy over Bellesiles's claims and evidence, the industrial innovation of using machines to produce firearms with interchangeable parts, achieved in the middle decades of the nineteenth century, greatly expanded production, reduced prices, and surely increased the incidence of ownership. For the production story, see generally Merritt Roe Smith, *Harper's Ferry and the New Technology: The Challenge of Change* (Ithaca, NY: Cornell University Press, 1977).

14. Laurel Thatcher Ulrich, *The Age of Homespun: Objects and Stories in the Creation of an American Myth* (New York: Knopf, 2001), 138.

15. Ruth Oldenziel, *Making Technology Masculine: Men, Women, and Modern Machines in America, 1870–1945* (Amsterdam: Amsterdam University Press, 1999), chap. 1; and Eric Schatzberg, "*Technik* Comes to America: Changing Meanings of *Technology* Before 1930," *Technology & Culture* 47, no. 3 (July 2006), 486–512.

Chapter 1. The Advent of Technology Consumption

1. Carlene E. Stephens, *On Time: How America Has Learned to Live by the Clock* (Boston: Little, Brown and Co., 2002), 50.

2. Ibid., 39; Michael O'Malley, *Keeping Watch: A History of American Time* (New York: Viking Press, 1990), 11.

3. Stephens, *On Time*, 99–100.

4. David A. Hounshell, *From the American System to Mass Production, 1800–1932: The Development of Manufacturing Technology in the United States* (Baltimore: Johns Hopkins University Press, 1984), 51–57.

5. Walter A. Friedman, *Birth of a Salesman: The Transformation of Selling in America* (Cambridge, MA: Harvard University Press, 2004), 20–21.

6. Joseph T. Rainer, "The 'Sharper' Image: Yankee Peddlers, Southern Consumers, and the Market Revolution," *Business and Economic History* 26, no. 1 (Fall 1997), 27–44; 30, 31; George Featherstonhaugh, quoted in Stephens, *On Time*, 85.

7. Rainer, "The 'Sharper' Image," 31.

8. Booker T. Washington, *Up from Slavery: An Autobiography* (1901; repr., Garden City, NY: Doubleday & Company, 1963), 81.

9. Based on observation of clocks in the collection of the NMAH.

10. Label reproduced in Kenneth D. Roberts, *The Contributions of Joseph Ives to Connecticut Clock Technology, 1810–1865* (Bristol, UK: Bond Press, 1970), 65.

11. Siegfried Gideon, *Mechanization Takes Command: A Contribution to Anonymous History* (New York: Oxford, 1948), 62–71.

12. Quoted in O'Malley, *Keeping Watch*, 67.

13. See, for example, Eli Terry Shelf Clock, c. 1817, catalog number 317,663, NMAH. Thanks to Carlene Stephens for enabling me to examine the clocks in the Smithsonian's collection and providing information and photographs.

14. Grace Rogers Cooper, *The Sewing Machine: Its Invention and Development*, 2nd. ed. (Washington, DC: Smithsonian Institution Press, 1976), 7–26.

15. Ibid., 28.

16. Ibid., p. 40, figure 37, "Table of Sewing Machine Statistics."

17. The "Little Worker" machine is discussed in Carter Bays, *The Encyclopedia of Early American Sewing Machines* (Columbia, SC, 1993), 118, 122; see also, White Sewing Machine Company, *Sayings Wise and Otherwise, with Some Bottom Facts in Black and White* (New York, n.d.), back cover, "White Sewing Machine and Manufacturing Co. Bridgeport, Conn.," WC.

18. Fannie A. Stall Alden, *Edith Burnham, or, How the Homestead Was Saved* (Middletown, CT: Pelton & King, 1876).

19. Ibid., 15–16.

20. Ibid., 17–20.

21. Thomas K. Ober to Singer Mfg. Co., Feb. 24, 1888, box 41, folder 7, SMCR.

22. "Instructions as to Making Lease Accounts and Prompt Collections of Same," July 4, 1888. box 32, folder 7, "Domestic Corr., Milwaukee, 1875–1888," SMCR.

23. J. H. Batton to Singer, Dec. 4, 1886, box 32, folder 6, "Domestic Correspon-

dence, Memphis, 1876" (folder either wrongly dated or letter erroneously filed), SMCR.

24. Although Singer's New York City branch was leasing machines at a dollar a week, in one week in 1896 it leased 7,295 units while reporting 696 repossessions, with some 4,247 accounts listed as "delinquent." "One Dollar Weekly Lease Report, Week Ending June 25, 1896," box 39, folder 3, "New York City, Weekly Lease Reports, 1895–96," SMCR.

25. *The Great Trial or Contest for Superiority between Sewing Machines* (Hartford, CT: Hutchings Printing House, 1869). This pamphlet about the trial, held under the direction of the Maryland Institute of Baltimore, contained excerpts from the press about the competition; in box 4, folder: "Wood Sewing Machine Co.," WC; for urban sewing machine sales emporia, see Hounshell, *From the American System,* 84–85. For agricultural competitions, see Joseph and Frances Gies, *The Ingenious Yankees: The Men, Ideas, and Machines That Transformed a Nation, 1776–1876* (New York: Thomas Crowell, 1976), 192; and Brooke Hindle and Steven Lubar, *Engines of Change: The American Industrial Revolution, 1790–1860* (Washington, DC: Smithsonian, 1986), 95. Harvesting and other agricultural machinery were also complex, and these too were sold with printed instructions that must have puzzled purchasers, but such machines were "producer" rather than consumer goods and thus outside this study.

26. Amy Isaacs, "Secret Weapon of the Union: The Sewing Machine and the Civil War." Undergraduate honors essay, Department of History, Stanford University, 2005.

27. F. Gilbart, Newark, to Singer Manufacturing Co., New York, Mar. 15, 1888, reporting on comments of an anonymous "Mrs. G," box 41, folder 3, WSHS.

28. *Directions and Instructions for Buying and Using the Wilson Shuttle Sewing Machine* (Cleveland, OH: Wilson Sewing Machine Co., n.d.), 48, WC. In citing owner's manuals here and elsewhere in the notes, I give the library or archive where I consulted an item. Although these materials are published, few libraries have acquired such ephemera and therefore they are exceedingly hard to find.

29. *Grover & Baker's Improved Lock Stitch Sewing Machine, No. 9,* "Sewing Machines," box 1, folder: "Grover & Baker Sewing Machine Co.," "Interesting Facts," unpaginated, WC.

30. *Directions and Instructions for . . . Wilson Shuttle,* 34–35, "Sewing Machines," box 1, folder: "Grover & Baker Sewing Machine Co.," "Interesting Facts," unpaginated, WC.

31. *Willcox & Gibbs Silent Family Sewing Machine* (Philadelphia, 1869), 68, 69; "Sewing Machines," box 1, folder: "Willcox & Gibbs Sewing Machine Company, Jersey City, New Jersey," unpaginated, WC.

32. Ibid., 70.

33. Thos. K. Ober, agent with Philadelphia office, to New York headquarters, Sept. 15, 1879, and May 11, 1880, box 41, folder 6, "Philadelphia, 1874–86," WSHS.

34. "To the Ladies and the Public Generally," from S. A. Wright, dealer in and repairer of all makes of sewing machines at lowest prices (n.p., n.d.), box 52, folder 1, WSHS; see also Cooper, *The Sewing Machine*, 159, 161.

35. The Singer Manufacturing Co., *Canvassers' Manual of Instructions* (Atlanta, n.d.), 14, box 109, folder 5, WSHS.

36. Ibid.

37. Handbill, n.d., box 52, folder 1, WSHS.

38. F. Gilbart to Singer Manufacturing Company, Mar. 15, 1888, box 41, folder 3, "Newark, 1876," WSHS; Gilbart does not use the saleswoman's actual name, he explains, "as she would likely object."

39. "Arrival of the No. 8 Wheeler & Wilson," c. 1885, Forbes and Company, lithographers, illustrated in Robert Jay, *The Trade Card in Nineteenth-Century America* (Columbia: University of Missouri Press, 1987), 83.

40. Alden, *Edith Burnham*, 22.

41. Instruction *Book for the Howe Sewing Machine, Step Feed* (New York: The Howe Machine Co., 1867), 7.

42. David H. Shayt, "Footpower in the Developing World: An American Precedent?" *Industrial Heritage '84 Proceedings*, vol. 2 (Washington, DC: Society for Industrial Archaeology, 1984), 44–51.

43. *A Home Scene, or, Mr. Aston's First Evening with Grover & Baker's Celebrated Family Sewing Machine Containing Directions for Using* (New York: Grover & Baker Sewing Machine Company, n.d. [c. 1859]).

44. I'd like to thank Maggie Dennis and Heather Paisley-Jones, of the Smithsonian—Maggie for suggesting that I try using a treadle machine, and Heather for loaning me the education department's old Singer treadle version. As a novice on the treadle, I quickly learned how difficult it was to control both the mechanism's direction and speed!

45. R. John Brockmann, *From Millwrights to Shipwrights to the Twenty-first Century: Explorations in a History of Technical Communication in the United States*. Written Language, edited by Marcia Farr (Cresskill, NJ: Hampton Press 1998), chap. 5.

46. New York: Domestic Sewing Machine Co., 1880, NMAH; Howe Machine Company, New York, *The Howe Sewing Machine Instructor* (1867–86).

47. Wheeler and Wilson Sewing Machine Manufacturing Company, "To the Purchaser" preface by distributor, box 4, file 4/18, WC.

48. Box 32, folder 7, "Domestic Correspondence, Milwaukee, 1875–1888," WSHS.

49. Brockmann, *From Millwrights to Shipwrights*, 160–61; illustrations in some

sewing machine manuals could be found on 29 percent of the instructional pages. Table 5.2, 168.

50. *Book of Directions for Wheeler & Wilson Home Sewing Machine* (Syracuse: J. G. Ayres, n.d.), 1, "Sewing Machines," box 4, file 4/18, "Wheeler & Wilson Sewing Machine Manufacturing Co.," WC; *The "Household" Sewing Machine Instructor* (Providence, n.d.), 3, WC.

51. *Description and Directions for Using the Willcox & Gibbs Sewing Machine* (n.p., n.d.), 2–4, "Sewing Machines," box 4, file 4/18, "Wheeler & Wilson Sewing Machine Manufacturing Co.," WC.

52. I used the phrase "hands-on illustration" first in my essay "Educating the Enthusiast: Print and the Popularization of Technical Knowledge," in *Possible Dreams* (Dearborn, MI: Henry Ford Museum, 1992), 19–33.

53. *Directions for Operating the New Davis Sewing Machine* (Watertown, NY: Times and Reformer Printing House, 1876), 14.

54. Caroline G. Stuart to Singer Manufacturing Company, July 12, 1864, Micro 2014, Reel 16 P92–9260, WSHS.

55. *Instruction Book for the Howe Sewing Machine, Step Feed* (New York: Howe Machine Co., 1871), 13, "Sewing Machines," box 1, folder 26, WC.

56. *Directions for Using the New Willcox & Gibbs Silent Sewing Machine with Automatic Tension* (New York: Willcox & Gibbs Sewing Machine Co., 1875), 15, "Sewing Machines," box 4, file 4/27, WC.

57. Sewing machine tools would have arrived in many an urban middle- or upper-middle-class home well before male members of the household might have acquired tools for a woodworking hobby. See Steven M. Gelber, "Do-It-Yourself: Constructing, Repairing and Maintaining Domestic Masculinity," *American Quarterly* 49, no. 1 (1997), 66–112; and Carolyn M. Goldstein, *Do It Yourself: Home Improvement in Twentieth-Century America* (Princeton, NJ: Princeton Architectural Press, 1998), 16–17.

58. The only objects put together with screws and bolts likely to be found in nineteenth-century American homes were pianos and parlor organs, increasingly popular among the middle classes. Manufacturers of such instruments, however, neither expected nor desired that buyers would touch those fasteners and did not provide them with tools to do so. Thanks to Gary Sturm, curator of musical instruments at the Smithsonian's NMAH, for this information.

59. *Directions for Using the New Willcox & Gibbs Silent Sewing Machine with Automatic Tension*, 11.

60. Advertising flyer for *Grover & Baker's Improved Lock Stitch Sewing Machine, No. 9* (n.p., n.d.), WC.

61. Alden, *Edith Burnham*, 21.

62. *Instruction Book for the Howe Sewing Machine Step Feed* (New York: Howe Manufacturing Co., 1867), 19.

63. The effort of sewing machine makers to control consumer interaction with their products to minimize frustrations and complaints paralleled management's attempts to control their own workers, the presumption in both cases being that management knew best. For the management story, see JoAnn Yates, *Control through Communications: The Rise of System in American Management* (Baltimore: Johns Hopkins University Press, 1989).

64. See, for example, *Instruction Book for the Howe Sewing Machine Step Feed* (New York: n.p., 1871), WC.

65. Until the 1880s, Singer machines were manufactured using the European system of hand fitting, and so presumably a consumer who ordered a spare part might find that it did not fit; see, Hounshell, *From the American System*, 85–91.

66. Mrs. C. D. Officer to Singer Manufacturing Company, July 6, 1864, Micro 2014, Reel 16 P92–9260, WSHS.

67. *The Home Sewing Machine Instructor* (New York: Howe Machine Co., 1867–86), 17, accessed at and printed from http://www.sil.si.edu/DigitalCollections/Trade-Literature/Sewing-Machines/SIL10-870-5.htm on Apr. 18, 2003, although that URL no longer works.

68. *Directions for Using the "Domestic" Sewing Machine* (New York: n.p., 1872), 7, HL; *Directions for Setting Up and Using the Union Button-Hole Sewing-Machine* (Boston, n.d.), 4–5, WC.

69. *Instruction Book for the Howe Sewing Machine Step Feed*, 19.

70. *Book of Directions for Using the Wheeler & Wilson Sewing Machine*, 14.

71. Ibid., 8; *Description and Directions for Using the Willcox & Gibbs Sewing Machine* (New York, n.d.), 4; *Directions for Operating the New Davis Sewing Machine* (Watertown, NY, 1876), 1.

72. *Directions and Instructions for Buying and Using the Wilson Shuttle Sewing Machine* (Cleveland, n.d.), p. 19.

73. I use the word *imagine* purposefully, for even with patterns users often found it difficult to produce garments that fit as well as ready-made clothing, increasingly available for men starting in the 1830s and for women after the Civil War. See Claudia B. Kidwell and Margaret C. Christman, *Suiting Everyone: The Democratization of Clothing in America* (Washington, DC: Smithsonian Institution Press, 1974), 53; and Jenna Weissman Joselit, *A Perfect Fit: Clothes, Character, and the Promise of America* (New York: Henry Holt and Co., 2001), 12–14.

74. *Sewing Machine News* (Aug. 1885), quoted in Brockmann, *From Millwrights to Shipwrights*, 187–88.

75. For the variety of mechanical devices entering American homes in the

period, see Harvey Green, *The Light of the Home: An Intimate View of the Lives of Women in Victorian America* (New York: Pantheon, 1983).

76. Gary Tobin, "The Bicycle Boom of the 1890s: The Development of Private Transportation and the Birth of the Modern Tourist," *Journal of Popular Culture* 7 (Spring 1974), 838–49.

77. Frances E. Willard, *How I Learned to Ride the Bicycle—Reflections of an Influential 19th Century Woman* (Sunnyvale, CA: Fair Oaks Publishing, 1991; reprint edition of *A Wheel within a Wheel,* 1895), 75.

78. Maria E. Ward, *Bicycling for Ladies* (New York: Brentano's, 1896), x.

79. Sercombe-Bolt Mfg. Co., *Catalog for Telegram Bicycles* (Milwaukee, 1893), 22, "Bicycles," box 3, WC.

80. H. C. Cushing Jr., compiler, *The Bicycle Hand-book for 1897: A Collection of Useful and Practical Information for Wheelmen and Wheelwomen from the Best Authorities* (New York, 1897), 17.

81. Claude Fischer, *America Calling: A Social History of the Telephone to 1940* (Berkeley: University of California Press, 1992), 123 and chap. 5 in general.

82. Carolyn Marvin, *When Old Technologies Were New: Thinking about Communications in the Late Nineteenth Century* (New York: Oxford University Press, 1988), chap. 2.

83. Of a thousand automobiles listed by maker in the 1904 Detroit city automobile registry, forty-four, or nearly 5 percent, were owner-built. David Kirsch, conference paper at the 2000 meeting of the Society for the History of Technology.

Chapter 2. Buying an Automobile

1. Smith Hempstone and Donald H. Berkebile, *The Smithsonian Collections of Automobiles and Motorcycles* (Washington, DC: Smithsonian, 1968), 70.

2. James J. Flink, *America Adopts the Automobile, 1895–1910* (Cambridge, MA: MIT Press, 1970), 36–37.

3. An early Ford investor, banker John S. Gray, felt this way as late as 1905 and would have sold his shares in the auto company but didn't want to pass on the risk of failure to others. Douglas Brinkley, *Wheels for the World: Henry Ford, His Company, and a Century of Progress, 1903–2003* (New York: Viking, 2003), 70.

4. Flink, *America Adopts the Automobile,* 34.

5. Richard and Nancy Fraser, *A History of Maine Built Automobiles, 1834–1934* (East Poland: R & N Fraser, 1991), 163.

6. See generally, C. P. Berry, *A Treatise on the Law Relating to Automobiles* (Chicago: Callaghan & Company, 1909), chaps. 5 and 6.

7. Michael L. Berger, *The Devil Wagon in God's Country: The Automobile and Social Change in Rural America, 1893–1929* (Hamden, CT: Archon Books, 1979).

8. Clay McShane, *Down the Asphalt Path: The Automobile and the American City* (New York: Columbia University Press, 1994), 176–77.

9. Beverly Rae Kimes, *Pioneers, Engineers, and Scoundrels: The Dawn of the Automobile in America* (Warrendale, PA: SAE International, 2005), 190.

10. *New York Times*, Mar. 4, 1906.

11. James J. Flink, *The Automobile Age* (Cambridge, MA: MIT Press, 1988), 29, 55; and Jean-Pierre Bardou et al., *The Automobile Revolution: The Impact of an Industry* (Chapel Hill: University of North Carolina Press, 1982), 15.

12. Rudi Volti, "A Century of Automobility," *Technology and Culture* 37, no. 4 (Oct. 1996), 663; "Firsts" are often disputed, and such judgments often turn on the definition of words like *large* as used in the claim by writer Robert B. Jackson that the Stanley brothers' 1899 building of "a hundred or so cars" was "the first time that a large number of identical automobiles had been offered for sale in the United States." *The Steam Cars of the Stanley Twins* (New York: Henry Z. Walck, Inc., 1969), 17.

13. U.S. Department of Commerce, Bureau of the Census, *Historical Statistics of the United States from Colonial Times to 1970*, part 1 (Washington, DC: USGPO, 1975), 716.

14. See generally, George H. Simmons, M.D., ed., "The Automobile for the Physician: As a Part of His Professional Equipment and as a Means of Pleasure and Recreation," *Journal of the American Medical Association* 54, no. 15 (Jan.–June 1910), 1245–76; and the series of short pieces by physicians published as "Automobiles for Physicians' Use," *Journal of the American Medical Association* (Apr. 21, 1906), 1172–1211, copy in subject file, "Physicians," in Div. of Transportation, NMAH.

15. A. B. Filson Young, *The Complete Motorist* (London: Methuen & Co. 1904), see especially chap. 7, "The Selection of a Motor-Car," and 161.

16. Winton Motor Carriage Co., *The Vertical Four-Cylinder Winton of 1905* (Cleveland, n.d.), 11, AACA; Terry B. Dunham and Lawrence R. Gustin with the Staff of Automobile Quarterly, *The Buick: A Complete History*, 3rd. ed. (Kutztown, PA: Kutztown Publishing Co., 1987), 53.

17. "Selecting an Automobile," *Automobile* 25, no. 9 (Aug. 31, 1911), 356.

18. The editors claimed to have "established" their department "Lessons of the Road" with the very object of encouraging this feeling of mutuality between maker and user, and accentuating the necessity of practical road experience for the satisfactory progress of the industry. *Horseless Age* 7 no. 11 (Dec. 12, 1900), 11.

19. James Hamilton, "4,000 Miles over the Hills of Old Vermont," *Horseless Age* 13, no. 4 (Jan. 27, 1904), 100.

20. *Leslie's Monthly*, May 1905, accession 657, box 9, "Winton," FA.

21. *1904 Winton, Advance Information* (Cleveland: The Winton Motor Carriage Co., Oct. 28, 1903), unpaginated (thanks to Roger C. Allison for copy).

22. Loyd A. Thomas, "How to Control a Motor Car in Emergencies," *McClure's* 24 (Jan. 1905), 49–50.

23. Ibid.

24. Young, *The Complete Motorist*, 164, 161.

25. See example where the editor says the question of the efficacy of air-cooled versus water-cooled engines "cannot be fairly treated in a 'trade' journal if the answer is to be of a specific character," *Automobile* 20, no. 5 (Feb. 4, 1909), 249; Brinkley notes that "automobile historians looking for road tests of 1920–1950 American cars have to turn to Britain's *The Automobile* in hopes of finding reviews and statistics" (*Wheels for the World*, 555).

26. Members paid a license fee for the right to use Selden's 1899 patent on the gasoline engine. Although some eighty automobile manufacturers received licenses, Henry Ford refused to recognize the "monopoly" and contested it in court. A ruling in 1911 finally upheld the patent on narrow grounds, concluding that it only applied to the Brayton two-cycle engine, an early form all but obsolete, as Ford and most other automakers had exploited the Otto four-cycle internal combustion design. The ruling caused the instant collapse of the ALAM. Flink, *America Adopts the Automobile*, 318–25.

27. ALAM, *Handbook of Gasoline Automobiles* (New York, 1905), 60–63, 118–21.

28. *Horseless Age* 21, no. 9 (Feb. 26, 1908), 125.

29. See *Automobile* 21, no. 25 (Dec. 16, 1909), 1049, for an unusual example of an automotive periodical's recommending particular vehicles, which occurred when a North Dakota reader wrote to the publication, saying he was "looking for a 1910 two-seated roadster, about 24 to 30 horsepower, weight about 1,500 pounds, road clearance say 10 to 12 inches, body set low, 110 to 112 in. wheelbase, selective type of gear shifting, cost from $1,000 to $1,500. Can you put me into communication with the manufacturer of such a car?" The editor listed five cars by name that met the writer's requirements, all drawn from its own reporting "at the time of the New York show," along with a second list of eighteen that "seem to meet most of your requirements." The editors avoided making any judgment about quality, however, leaving the prospect to whittle down the list on his own.

30. Young, *The Complete Motorist*, 161, 164–65.

31. Franklin Pierce, *Motor Car Anatomy, Being a Book of Valuable Information for the Prospective Purchaser of an Automobile* (n.p., 1912), 75, JCL.

32. William E. Sharp, "The Choice and Care of an Automobile," *Journal of the American Medical Association* 54 (Jan.–June 1910), 1248.

33. Staebler to Mr. Tannehill of the Dodge Motor Car Company, Apr. 20, 1915, Michael Staebler Papers, box 10, letterpress book vol. 6, June 13, 1912, to Sept. 24,

1913, BL. Relying on such visual representations could be misleading, however, as artists had long exaggerated the scale of manufacturing premises in lithographs done for calendars, stationery, and other illustrations commissioned by businesses.

34. Herbert W. Manahan, M.D., "The Steam and the Gasoline Car for the Physician's Use," in "The Automobile for the Physician," special issue, *JAMA* 54, no. 15 (Apr. 9, 1910), 1247.

35. C. H. Bryan, M.D., "Details from Experience," *Journal of the American Medical Association* (Apr. 21, 1906), 1173 (copy in NMAH Transportation Division's vertical files).

36. See "The Accessible Car," *Automobile* 25, no. 21 (Nov. 23, 1911), 916; also "Accessibility Reduces Cost of Repairs," *Automobile* 26, no. 1 (Jan. 4, 1912), 66–72.

37. The American right-of-way rules derived from the fact that the driver of a horse-drawn vehicle always sat on the right side of the seat, holding the reins in his left hand leaving his right free to work the whip, which wouldn't endanger somebody sitting next to him on his left. Additionally, in that position the driver could better monitor the deep ditches on the right side of the roads. This practice helps explain why American cars up to about 1915 mostly had right-hand steering; Peter Kincaid, *The Rule of the Road: An International Guide to History and Practice* (New York: Greenwood Press, 1986), 27.

38. H. Wieand Bowman and Robert J. Gottlieb, *Classic Cars and Antiques*, Trend Book 111 (Los Angeles: Trend Books, 1953), 22, 52; Automobile Quarterly, *General Motors: The First 75 Years* (New York: Crown Publishers, 1983), 20–30.

39. "Right or Left Control—Which?" *Automobile* 27, no. 5 (Aug. 1, 1912), 209–13, 241; see also follow-up letters in no. 6 (Aug. 8, 1912), 228, 269; no. 9 (Aug. 29, 1912), 438–39.

40. Similarly, some left-steer vehicles located the control levers to the driver's left. This and the reverse arrangement, right-hand steering with levers to the right, all but made it impossible for anybody to get in or out on the driver's side of the car. *Automobile* 27, no. 6 (Aug. 8, 1912), 210.

41. Sound engineering considerations also supported left-steering: with the engine's torque already placing greater load on the right wheels of the car due to the crowned design of many roads, which were higher in the center; placing the driver on the left side and moving the control levers to a central position would favorably reduce the load over the right wheels. In addition, left-hand steering would permit drivers, most of whom were right-handed, to shift with their right hands. *Automobile* 27, no. 6 (Aug. 8, 1912), letter to editor from J. G. Perrin, an engineer with the Lozier Motor Co. (Aug. 8, 1912), 228, 269.

42. *Automobile* 27, no. 5, 212.

43. Gijs Mom, *The Electric Vehicle: Technology and Expectations in the Automobile Age* (Baltimore: Johns Hopkins University Press, 2004), 57.

44. *Horseless Age* 21, no. 12 (Mar. 18, 1908), 303; the editors of *Motor Age* similarly edited readers' letters to protect car manufacturers but also complained of automakers who didn't provide information on their products to the journal but always expected of any coverage "that it will be something 'nice.'" *Motor Age* 6, no. 21 (Nov. 24, 1904), 4; one periodical did not expurgate the names of marques in articles; see "A Doctor's View, His Experience with a Car and Why He Likes It," by an unnamed Exeter, NH, physician, dealing with his "second hand Maxwell runabout," that was printed "somewhat abridged as to non-essentials." *Automobile Dealer and Repairer* 9, no. 6 (Aug. 1910), 946.

45. James Hamilton, "4,000 Miles over the Hills of Old Vermont," *Horseless Age* 13, no. 4 (Jan. 27, 1904), 100–101.

46. *Horseless Age* 15, no. 5 (Feb. 1, 1905), 155.

47. *Horseless Age* 15, no. 23 (June 7, 1905), 631. See also Jackson, *The Steam Cars of the Stanley Twins,* 46.

48. A. H. Hill, "The Advantages and Disadvantages of the Steam Car," in "The Automobile for the Physician," special issue, *JAMA* 54, no. 15 (Apr. 9, 1910), 1269.

49. Beverly Rae Kimes and Henry Austin Clark Jr., *Standard Catalog of American Cars, 1805–1942* (Iola, WI: Krause Publications, 1985), 819, 1298, 1420, 1455; John Bentley, *Oldtime Steam Cars* (Greenwich, CT: Fawcett Books, 1953), 108, 114.

50. Michael B. Schiffer, *Taking Charge: The Electric Automobile in America* (Washington, DC: Smithsonian, 1994), 95, 142; Mom, *The Electric Vehicle,* 146, 157.

51. Discussion by Dr. H. R. Boettcher of address by William E. Sharp, "The Choice and Care of an Automobile" at the Chicago Medical society's Automobile Clinic, Feb. 1, 1910, in "The Automobile for the Physician," special issue, *JAMA* 54, no. 15 (Apr. 9, 1910), 1250–51.

52. Mom, *The Electric Vehicle,* 105; see also, on infrastructure, David A. Kirsch, *The Electric Vehicle and the Burden of History* (New Brunswick, NJ: Rutgers University Press, 2000), 167–88.

53. Students of the technology-society relationship often speak of "technology choice" in describing the collective decisions of consumers, along with the manipulations of producers and others, that result in one variant of a technology winning out, so to speak, over another. For an application of the concept, see Joy Parr, "What Makes Washday Less Blue? Gender, Nation, and Technology Choice in Postwar Canada," *Technology and Culture* 38, no. 1 (Jan. 1997), 153–86. Also see the methodological approach developed by sociologists and historians of technology termed "social construction," outlined and exemplified by the volume edited by Wiebe E. Bijker, Thomas P. Hughes, and Trevor Pinch, *The Social Construction of Technological Systems* (Cambridge, MA: MIT Press, 1987).

54. Virginia Scharff, *Taking the Wheel: Women and the Coming of the Motor Age* (New York: Free Press, 1991), 37, 45; Claudy quote, 41.

55. Mom, *The Electric Vehicle*, 107.

56. Baker Electric merged with Rauch & Lang, another producer, in 1915, and Waverley ended production in 1916; Detroit Electric continued the production of electrics until at least 1938. Kimes and Clark, *Standard Catalog of American Cars, 1805–1942*, 94, 405, 1440.

57. Allan Nevins and Frank Ernest Hill, *Ford: The Times, the Man, the Company* (New York: Charles Scribner's & Sons, 1954), 388.

58. See generally, Robert Casey, *The Model T: A Centennial History* (Baltimore: Johns Hopkins University Press), 2008.

59. "Right or Left Control—Which?" 210.

60. The Type 51, introduced in 1915, was the first left-hand drive car made by Cadillac, although the company continued to offer right-hand drive as an option; Walter M. P. McCall, *80 Years of Cadillac LaSalle* (Sarasota, FL: Crestline Publishing, 1982), 57–60.

61. *The Locomobile Book* (Apr. 1915), 13, box 8, folder 32, WC.

62. For Packard see *Town & Country* advertisement, Oct. 12, 1912, 14; Locomobile, *Town & Country* advertisement, June 28, 1913, 2, box 8, folder 31, "The Locomobile Company of America," WC. Pierce-Arrow continued to have right-hand drive until 1920 as an option.

63. Winton Motor Carriage Co. to Charles Henry, Apr. 22, 1905, Charles Roswell Henry Papers, File: "Papers 1905," BL.

64. Winton, *1904 Winton* (Cleveland, n.d.), unpaginated.

65. Floyd Clymer, *Henry's Wonderful Model T, 1908–1927* (New York: Bonanza Books, 1955), 159.

66. Harold B. Chase, *Auto-Biography: Recollections of a Pioneer Motorist, 1896–1911* (New York: Pageant Press, Inc., 1955), 78; for a Ford Motor Company defense of the relatively minimally equipped Model T as "completely equipped," see Don C. Prentiss, *Ford Products and Their Sale: A Manual for Ford Salesmen and Dealers in Six Books* (Detroit: The Franklin Press, 1923), 399–400, HBL.

67. Looking back on the bewildering choices that rapidly came to be "taken for granted," M. M. Musselman wrote, "Today one seldom questions the style of a clutch, the mechanism of the brakes, the type of engine. There is no longer a choice to be made between right and left drive, three or four cylinders, gear or friction transmission, chain or shaft drive, underhung or overhung chassis, rear or side doors, two or four cycles, self-generating acetylene or Presto-lite illumination." *Get a Horse: The Story of the Automobile in America* (Philadelphia: J. B. Lippincott Co., 1950), 22.

68. *Ford Times* 5, no. 6 (Apr.–May 1912), 205.

69. Such visual practices were codified in John F. Brennan, *Automobile Identification* (New York: Scientific American Publishing Company, 1924).

70. NMAH, Da Vincent Collection, "Transportation," box 13, folder x, "The Packard and the Ford," words by Harold R. Atterbridge, music by Harry Carroll, 1915.

71. Samuel Stanley to Henry Ford, May 22, 1926, accession 94, box 168, folder: "Complaints, General, 1926," FA.

72. Box 10, "Automobile Vocals," DV.

73. Ibid., "TE-Z."

74. Statistics from Harold Katz, *The Decline of Competition in the Automobile Industry, 1920–1940* (New York: Arno, 1977), 41, cited in Richard S. Tedlow, *New and Improved: The Story of Mass Marketing in America* (New York: Basic Books, 1990), 155.

75. For the history of the residential garage, see Drummond Buckley, "A Garage in the House," in *The Car and the City: The Automobile, the Built Environment, and Daily Urban Life*, eds. Martin Wachs and Margaret Crawford (Ann Arbor: University of Michigan Press, 1991), 124–40; J. B. Jackson, "The Domestication of the Garage," *Landscape* 20, no. 2 (1976), 10–19; and Folke T. Kihlstedt, "The Automobile and the Transformation of the American House, 1910–1935," in *The Automobile and American Culture*, eds. David L. Lewis and Laurence Goldstein (Ann Arbor: University of Michigan Press, 1980), 160–75.

76. For traffic signals and control, see Clay McShane, "The Origins and Globalization of Traffic Control Signals," *Journal of Urban History* 25, no. 3 (Mar. 1999), 381–83; Peter D. Norton, *Fighting Traffic: The Dawn of the Motor Age in the American City* (Cambridge, MA: MIT Press, 2008), 54–63.

77. A good overview of the impact of mass automobility on the built environment is Chester H. Liebs, *Main Street to Miracle Mile: American Roadside Architecture* (New York: New York Graphic Society, 1985).

78. Steve Gelber, *Horse Trading in the Age of Cars: Men in the Marketplace* (Baltimore: Johns Hopkins University Press, 2008), 44–46.

79. James M. Rubenstein, *Making and Selling Cars: Innovation and Change in the U.S. Automotive Industry* (Baltimore: Johns Hopkins University Press, 2001), 253; Kimes and Clark, *Standard Catalog of American Cars, 1805–1942*, 185.

80. Flink, *America Adopts the Automobile*, 49.

81. Invoice of Steele to Henry, June 1, 1905, folder: "Correspondence, 1905," CRH.

82. Walter Dickenson of Glen Falls, NY, wrote the Ford Motor Company requesting a catalog of 1905 models, and the automaker in response wrote offering him a 20 percent discount as Ford is "not represented in your locality at the present time." Feb. 21, 1905, accession 64, FA.

83. The Staebler Papers, *Michigan Historical Collections Bulletin* 6 (Dec. 1953), 18–20, BL.

84. Staebler to United States Motor Co., NYC, June 20, 1911, letterpress book, May 18, 1911, to June 14, 1912, BL.

85. Staebler to Apperson Bros. Automobile Co., Kokomo, IN, July 8, 1911; letterpress books, vol. 1, May 3, 1904, to Feb. 12, 1907, Staebler to Pope Motor Car Co., Indianapolis (re: Waverly); vol. 7, Sept. 24, 1913 to May 15, 1915, Staebler to Dodge Bros., May 25, Aug. 5, 1914 (about Dodge agency), BL.

86. David L. Lewis, *The Public Image of Henry Ford* (1976), 40; Allan Nevins and Frank Ernest Hill, *Ford: Expansion and Challenge, 1915–1933*, vol. 2 (New York: Charles Scribner's Sons, 1957).

87. J. C. White, "Report on Bradford Bros.," Nov. 25, 1919, accession 76, box 88, "Louisville Branch," FA.

88. In 1919, the White Lake Garage in Whitehall, MI, was selling not only Fords but also the Dort, Reo, Chevrolet, and Oldsmobile automobiles along with Moline tractors; see "Roadman Report," Apr. 17, 1919, accession 76, box 56, FA.

89. *R. L. Polk & Co.'s Alpena City and County Directory, 1927–1928*, microfilm, reel 2 (Detroit, 1927), 272–75.

90. John A. Jakle and Keith A. Sculle, *The Gas Station in America* (Baltimore: Johns Hopkins University Press, 1994), 203.

91. Corey T. Lesseig, *Automobility: Social Changes in the American South, 1909–1930* (New York: Routledge, 2001), 46.

92. Pat Soberanis, "The Grandest Auto Row," *Motorland/CSAA* (July–Aug. 1990), 35–36; a 1921 statistical compilation claimed 7,817 Ford dealers among 59,521 dealers of "all makes," a number that seems high, see Martin Tuttle, *Automotive Statistics* (Des Moines, IA: Motor List Co., 1921), 79, JCL; 1920 statistic from Flink, *The Automobile Age*, 70.

93. Jack H. Morgan, *Profitable Automobile Merchandising* (South Bend, IN: Studebaker Corp., 1934), 23; Harold G. Vatter argued that 1923 saw the transformation of the new car market into primarily a replacement market, which corresponded with the ascension of the auto industry to first place in U.S. manufacturing along with the founding of the last member of the Big Three, Chrysler. "The Closure of Entry in the American Automobile Industry," *Oxford Economic Papers*, n.s., 4, no. 3 (Oct. 1952), 218–19.

94. Murray Fahnestock, "Salesmen! Do You Know Why?" *Fordowner* 24, no. 6 (Mar. 1926), 60.

95. See, for example, *Canvassers' Manual of Instructions* (Atlanta: Singer Manufacturing Co., n.d.), Singer Manufacturing Co. Records, box 109, folder 5, WSHS; Joseph H. Crane, "How I Sell a National Cash Register" (Dayton: National Cash Register Co., 1887), typescript, from uncataloged materials in NCR Archive, Dayton, OH (thanks to Susan V. Spellman for this reference); *The Edison Retail Sales Laboratory* (1915), an irregular publication addressed to phonograph salesmen

(thanks to Emily S. Thompson for this reference); and *A Salesman's Manual of the Royal Typewriter, With Typewriter Notes and Comparisons* (New York: 1918), HBL.

96. The earliest automotive example I have seen is *The Chalmers Sales Manual: A Book of Suggestions for Chalmers Salesmen* (Detroit: Chalmers Motor Company, 1916), HBL.

97. *Handbook for Use of Oldsmobile Salesmen* (Lansing, MI: Oldsmobile Motor Works, n.d.), title page, HBL.

98. *The Model T Specialist* (n.p., 1925), 49, FA.

99. Dodge Motor Co., *Used Cars, Sales Ammunition No. 1* (Detroit, 1925), 22, WC.

100. *Graham-Paige Sales Manual* (Dearborn: Graham-Paige Motors, 1929), 1, 24–25, HL.

101. *Cylinders or Satisfaction,* Sales Ammunition Series no. 21 (Detroit, 1926), 7, 9–10.

102. Don C. Prentiss, *Ford Products and Their Sale: A Manual for Ford Salesmen and Dealers in Six Books,* vol. 5 (Detroit: The Franklin Press, 1923), 400–401, HBL.

103. *The Chalmers Sales Manual,* 7–8.

104. *Selling Chevrolets: A Book of General Information for Chevrolet Retail Salesmen* (Detroit: the Chevrolet Motor Company, 1926), 104–5.

105. According to Henry L. Dominguez, a bare chassis was "omnipresent in all of Ford's exhibits; primarily in branch offices, but occasionally in dealerships, too." *The Ford Agency: A Pictorial History* (Osceola, WI: Motorbooks, 1981), 49.

106. *Selling Chevrolets,* 110; "Selling Cars without Lifting the Hood," *Motor World* 38, no. 1 (Jan. 15, 1912), 18; I thank Steve McIntyre for this reference.

107. "Wiles of the Demonstrator," *Motor World* 10, no. 8 (May 18, 1905), 355.

108. Barney Oldfield, *Barney Oldfield's Book for the Motorist* (Boston: Maynard Co., 1918), 55–56.

109. *Chrysler Bulletin No. 5, Confidential Retail Salesman's Service* (n.p., Nov. 26, 1940), box 4, WC. I have not been able to determine when exactly the current practice emerged whereby dealers just turn over the keys of a car to an interested prospect, who then takes it for a test drive without the salesman.

110. "Has Special Room for Closing Contracts," *Motor World* 42, no. 1 (Jan. 6, 1915), 44.

111. *Selling Chevrolets,* 102, 108–9.

112. Ford Motor Company, *Sales Manual* (Dearborn, 1926–27), 44–45, accession 175, FA.

113. *Chrysler Bulletin No. 7, Confidential Retail Salesman's Service* (n.p., Dec. 30, 1940), unpaginated, box 4, WC.

114. Gary Cross, *An All-Consuming Century: Why Commercialism Won in Modern America* (New York: Columbia University Press, 2000), 29.

115. Guarantee Securities Company started financing dealer sales of Overland automobiles in 1915, and the demand was so great that the firm started financing all makes the following year; Casey, *The Model T*, 84.

116. "The Reminiscences of Mr. John H. Eagal, Sr." Typescript of interview (June 15, 1952), 52, FA.

117. *The Model T Specialist*, 53.

118. U. N. Versal, "Let Us Present the Young Woman Who Knows How to Sell Used Cars," *Ford Dealer and Service Field* (Nov. 1926), 56–58.

119. Rubenstein, *Making and Selling Cars*, 276; see also Martha L. Olney, *Buy Now, Pay Later: Advertising, Credit, and Consumer Durables in the 1920s* (Chapel Hill: University of North Carolina Press, 1991), 9, 33, 40, 92, and 95.

120. "Second Hand Cars," Feb. 23, 1916, service letter 115, accession 235, "Finance," box 39, FA.

121. Walter A. Friedman, *Birth of a Salesman: The Transformation of Selling in America* (Cambridge, MA: Harvard University Press, 2004), 219.

122. Quoted in Lesseig, *Automobility*, 38.

123. Paul G. Hoffman and James H. Greene, *Marketing Used Cars* (New York: Harper & Brothers, 1929), 14.

124. "Selling the Appraisal," *Chrysler and Plymouth Salesmanship* (Detroit, 1930), 1, FA.

125. Gelber, *Horse Trading in the Age of Cars*, 62.

126. Automobile registrations peaked in 1930 at just over 23 million and only exceeded that number in 1936; Bureau of the Census, *Historical Statistics*, 716.

127. William Ashdown, "Confessions of an Automobilist," *Atlantic Monthly* 135 (June 1925), 786–92.

128. Ibid.

Chapter 3. Running a Car

1. Harold W. Slauson, *Everyman's Guide to Motor Efficiency: Simplified Shortcuts to Maximum Mileage at Minimum Cost* (New York: Leslie-Judge Company, 1920), frontispiece.

2. Morris A. Hall, *Care and Operation of Automobiles: A Handbook on Driving, Road Troubles, and Home Repairs* (Chicago: American School of Correspondence, 1912), 71.

3. For references to "operators," see "Licensing—Vehicle Operators," vertical file, Division of Transportation, NMAH. For another typical use of *run*, see *Directions for the Operation, Care and Adjustment of Willys Knight Model 84 Automobiles* (Toledo: Willys-Overland Co., 1915): "In the chapters on operation we have set forth detailed instructions which will enable the motoring novice to master the

running of the car," 1. A more ambivalent synonym for *operate* or *run* was *manage*, applied occasionally to the direction of animals but also to the handling of automobiles; see, e.g., Gardner Dexter Hiscox *Horseless Vehicles, Automobiles: A Practical Treatise* (New York: N. W. Henley & Co., 1900), chap. 15, "The General Management of Motor Vehicles," 375, JCL.

4. James T. Sullivan, "Making Automobile Highways Safe," *Automobile* 30, no. 12 (Mar. 19, 1914), 635.

5. *Automobile* 22, no. 9 (Mar. 3, 1910), 478; no. 13 (Mar. 31, 1910), 643; 30, no. 16 (Apr. 16, 1914), 853; for a rather late use of some of these terms, see William Ashdown, "Confessions of an Automobilist," *Atlantic Monthly* 135 (June 1925).

6. Contemporaries ardently debated the suitability of the new coinages put forth to describe a person at the controls of a motor vehicle but preferred the name operator over " 'words of learned length' and foreign derivation" such as the French *automobilist*. Editorial, "Terminology," *Horseless Age* 4, no. 10 (June 7, 1899), 5–6; see also Editorial, "Driver or Motorman?" *Horseless Age* 4, no. 12 (June 21, 1899), 7; and Editorial, "More Verbal Coinages," which rejected the suggestion of Prof. C. M. Woodward of Washington University in St. Louis, who, loathing the word *automobile,* proposed "that the second half of the word 'automobile' be discarded, and the new vehicle be called an 'autom,' " prompting the editor to conclude, "of all the freak words that have been offered, . . . this is about the worst." *Horseless Age* 4, no. 14 (July 5, 1899), 6.

7. "Man, Horse, Automobile, and the Highways," *Automobile* 2 (Jan. 9, 1908), 38–40.

8. *A Study of the Oldsmobile: A Little Booklet for Users of the Oldsmobile Curved Dash Runabout* (Detroit: Olds Motor Works, n.d.), 32.

9. Thomas H. Russell, *Automobile Driving Self-taught: An Exhaustive Treatise on the Operation, Management, and Care of Motor Cars* (Chicago: Charles C. Thompson Co., 1909; reprint edition 1971), 55.

10. William B. Sandison, *Annual Automobile Repair Guide and Speed Laws: A Complete Manual on the Care and Repair of the Automobile* (New York: W. B. Sandison, 1910), 19.

11. Nash Motor Company, *The Nash Four Instruction Book* (Kenosha, WI, 1912), 10.

12. James R. Wright, *Auto Sense* (Trenton, MO: W. B. Rogers Printing, 1919), 39.

13. Julian Street, "Good and Bad Driving: The Pneumatic Peril of the Highway," *Collier's* 46, no. 16 (Jan. 7, 1911), section 2, 9, 13.

14. George Harris et al., *Audels Answers on Automobiles for Owners-Operators-Repairmen* (New York: Theodore Audel & Co., 1912), preface, n.p.

15. Wright, *Auto Sense,* 39.

16. Roscoe Sheller, *Me and the Model T* (Portland, OR: Thomas Binford, 1965), 58.

17. A. Frederick Collins, *Keeping Up with Your Motor Car, Written So That He Who Reads May Ride; Also for the Car Owner to Whom Money Is an Object* (New York: D. Appleton and Company, 1917), 14.

18. Letter from Predmore, quoted in Miles S. Amick, "The Cross Engine Franklin," part 2, *Antique Automobiles* 20, no. 1 (Spring 1956), p. 22.

19. Myron M. Stearns, "Learning to Drive," *Saturday Evening Post* 198, no. 51 (June 19, 1926), 26.

20. James J. Flink, *America Adopts the Automobile, 1895–1910* (Cambridge, MA: MIT Press, 1970), 227–30.

21. These statistics from the New York City YMCA in 1915 are suggestive of the mixed group of students taking the courses: "Owners and Prospective Owners: 364; Chauffeurs: 613; Salesmen: 36; Garage Managers: 22; Repairmen: 27; Not Specified: 272." I thank Kevin Borg for sharing with me these findings from his research in the YMCA archives.

22. Connecticut, for example, a pioneer in licensing automobile drivers, made an exam obligatory in 1913, but it appears to have been written-only, and the state did not care how one learned; Typescript, "40th Anniversary 1917–1957," Hartford, CT: CT Dept. of Motor Vehicles, June 1, 1957, 2–3, in NMAH Division of Transportation vertical files, folder: "Licensing Registration and Regulation of Autos." In 1931 it was "illegal to drive an automobile without an operator's license in twenty-four states and the District of Columbia," or, putting it the other way, in nearly half of the states there were still no licensing requirements whatsoever; see Herbert L. McCann and James O. Spearing, *Expert Driving* (Brooklyn: Transportation Engineering Service, Inc., 1931), 8–9.

23. Marjorie M. Sweet, "Woman Pioneer Her Own Auto Mechanic in Early Days," *Old Times News* 2, no. 3 (July 1944), 30.

24. Quoted in Eleanor Arnold, *Buggies and Bad Times,* Memories of Hoosier Homemakers no. 3 (Indiana Historical Society, 1985), 41.

25. I examine some of this literature at greater length in chapter 5.

26. Scholars have extensively debated the broader question, posed famously by Robert L. Heilbroner in the title of his essay "Do Machines Make History?" That is, to what extent are historical processes and developments, including the emergence of particular technologies, themselves determined by technology? *Technology & Culture* 8, no. 3 (July 1967), 335–45; see also the stimulating collection of essays on the subject in *Does Technology Drive History: The Dilemma of Technological Determinism,* ed. Merritt Roe Smith and Leo Marx (Cambridge, MA: MIT Press, 1994).

27. Because we are interested here in the operator's relationship to the machine

and its controls, this list subordinates or altogether omits the crucial role of good judgment behind the wheel.

28. Robert B. Jackson, *The Steam Cars of the Stanley Twins* (New York: Henry Z. Walck, Inc., 1969), 48–50.

29. American Technical Society, *Automobile Engineering*, vol. 1 (Chicago: American Technical Society, 1927), 276.

30. The curved dash Oldsmobile, for example, had its "starting crank" in such a position; *A Study of the Oldsmobile* (Detroit: Olds Motor Works, n.d.), 13, 20.

31. In 1929 Studebaker warned its mechanics against "the evils of heavy engine oil," which it viewed as, "no doubt a holdover from the old days" when, as engines warmed up, the oil lost its viscosity and effectiveness. "The modern oils will maintain a relatively even viscosity throughout the crankcase temperature range." *Studebaker Service*, 42–A (Oct. 1929), 12.

32. In 1908 "no less than 49.2 per cent of all automobile accidents" were cranking accidents, according to the International Association of Accident Underwriters, cited in Editorial, "Cranking Dangers," *Horseless Age* 22, no. 4 (July 22, 1908), 99; the percentage had dropped by 1912 to 37 percent; cf. Hall, *Care and Operation of Automobiles*, 58.

33. "Surgeons Find 'the Chauffeurs' Fracture,'" *Motor World* 22, no. 7 (Feb. 17, 1910), 516.

34. Russell, *Automobile Driving Self-taught*, 51.

35. Beverly Rae Kimes, *Pioneers, Engineers, and Scoundrels: The Dawn of the Automobile in America* (Warrendale, PA: SAE International, 2005), 364.

36. For related discussion, see *Audels Answers on Automobiles for Owners-Operators-Repairmen* (New York: Theo. Audel & Co., 1911), 234.

37. Beverly Rae Kimes and Henry Austin Clark Jr., *Standard Catalog of American Cars, 1805–1942* (Iola, WI: Krause Publications, 1985), 504.

38. See Stuart W. Leslie, *Boss Kettering* (New York: Columbia University Press, 1983), 45.

39. Virginia Scharff, *Taking the Wheel: Women and the Coming of the Motor Age* (New York: Free Press, 1991), 61.

40. Hilda Ward, *The Girl and the Motor* (Cincinnati: The Gas Engine Publishing Co., 1908). "The idea of electrical self-starting was continually circulating among automotive engineers looking for a break," Leslie, *Boss Kettering*, 45.

41. Flink, *America Adopts the Automobile*, 283.

42. Kimes and Clark, *Standard Catalog of American Cars*, 532.

43. The 1932 Lincoln, for example, had a hand crank in its tool kit, see *Book of Instruction Lincoln Motor Cars* (Detroit: Lincoln Motor Company, 1932), 74–75.

44. Interview with John Goldsmith, New York City, June 10, 1987.

45. According to a War Production Board decree, old cars "other than a care-

fully preserved antique" could be classed as junk, and thus targeted by scrap drives; *Antique Automobiles* 1, no. 1 (Jan. 1943), 17.

46. Clinton Edgar Woods, *The Electric Automobile: Its Construction, Care and Operation* (Chicago: H. S. Stone & Company, 1900), 107.

47. Victor Lougheed, *How to Drive an Automobile* (New York: Motor, 1908), 4–5.

48. Ward, *The Girl and the Motor,* 101–2.

49. Sigmund Krausz, *Krausz's ABC of Motoring: A Manual of Practical Information for Layman, Auto Novice and Motorist* (Chicago: Laird & Lee, 1906), 121.

50. Charles V. Milward, *Practical Automobile Hand Book for Owners, Operators, and Mechanicians* (Philadelphia: Hathaway & Bros., 1912), 31.

51. Slauson, *Everyman's Guide to Motor Efficiency,* 225–26.

52. Arch Brown, "McFarlan TV Series, Most Powerful American Car of 1926," quoting owner Dick Squires, *Special Interest Autos,* no. 115 (Feb. 1990), 39. The McFarlan was an exceptionally heavy, powerful, and poor-steering machine, but the general point characterizes many cars of the 1920s and earlier.

53. "Does Driving a Car Become Automatic?" *Journal of the American Medical Society* 46, no. 14 (Apr. 9, 1910), 1276.

54. H. Clifford Brokaw and Charles A. Starr, *Putnam's Automobile Handbook: The Care and Management of the Modern Motor Car* (New York: G. P. Putnam's Sons, 1918), chap. 22, 137–42.

55. "It took great strength, skill and patience to manipulate early gearboxes," according to historian Philip G. Gott, *Changing Gears: The Development of the Automotive Transmission* (Warrendale, PA: Society of Automotive Engineers, 1991), i.

56. "Manipulation of the Clutch," *Motor World* 10, no. 5 (Apr. 27, 1905), 263.

57. Russell, *Automobile Driving Self-taught,* 37.

58. By the latter half of the twenties, double-clutching was becoming passé, a "collegiate trick," some called it, presumably because students drove older jalopies that still required the technique; "When the 'Expert' Pulls a Boner," *Literary Digest* 91, no. 10 (Dec. 4, 1926), 80.

59. Michael Lamm, "Model A: The Birth of Ford's Interim Car," *Special-Interest Autos* 4, no. 18 (Aug.–Oct. 1973), 56.

60. Robert Casey, *The Model T: A Centennial History* (Baltimore: Johns Hopkins University Press, 2008), 98.

61. Victor W. Pagé, *The Model T Ford Car* (New York: Norman W. Henley Publishing Co., 1926), 110.

62. E. B. White, "Farewell My Lovely! (An Aging Male Kisses an Old Flame Good-Bye, circa 1936)," *Essays of E. B. White* (New York: Harper & Row, 1977), 165.

63. Ibid., 163.

64. Victor W. Pagé, *How to Run an Automobile* (New York: Norman W. Henley, 1917), 105, 109–10, 125, 135.

65. Editorial, "Methods of Auto-Control," *Automobile* 24, no. 1 (Jan. 5, 1911), 25.

66. Avis Berman, *Rebels on Eighth Street: Juliana Force and the Whitney Museum of Art* (New York: Atheneum, 1990), 148; thanks to Alex Nemerov for this reference. On Stein, see Kimes, *Pioneers, Engineers, and Scoundrels*, 419.

67. Robert Sloss, *The Automobile: Its Selection, Care, and Use* (New York: Outing Publishing Co., 1910), 74.

68. Harold Whiting Slauson, "Saving the Car by Careful Driving," *Harper's Weekly* 56 (Mar. 2, 1912), 13.

69. Kimes and Clark, *Standard Catalog of American Cars*, 1361.

70. Gott, *Changing Gears*, 83; see also McCann and Spearing, *Expert Driving*, 42–46.

71. T. P. Newcomb and R. T. Spurr, *A Technical History of the Motor Car* (Bristol, UK: Adam Hilger, 1989), 247.

72. Ibid., 246.

73. Hall, *Care and Operation of Automobiles*, 53.

74. Making mechanically actuated brakes work on the front wheels of a car, which had to turn when steering, was a difficult engineering challenge, Pagé, *The Model T Ford Car*, 123.

75. CRH, Winton Motor Carriage Co. to Charles R. Henry, stamped "Received Aug. 8, 1905."

76. Roger B. Whitman, "Scorching in My Horseless Carriage," *Popular Science* (Sept. 1929), 134; see also "Skidding or Side-Sliding," chap. 32 of Charles E. Duryea and James E. Homans, *The Automobile Book: A Practical Treatise* (New York: Sturgis & Walton Company, 1916), 326–30.

77. Slauson, *Everyman's Guide to Motor Efficiency*, 220, says: "Use the brakes as little as possible."

78. H. S. Whiting, "Saving the Car by Careful Driving," *Scientific American* 114, no. 1 (Jan. 1, 1916), 18.

79. Eleanor Arnold, ed., *Buggies and Bad Times*, Indiana Extension Homemakers Association, Hoosier Homemakers through the Years Oral History Project (n.p., 1985), 39, copy in author's possession.

80. Walter Cronley, "My Wonderful Years with the Automobile," *Automotive Old Timers News* 3, no. 3 (July–Sept. 1969), 11.

81. "The Motor More Deadly Than War," *Literary Digest* 94 (Aug. 27, 1927), 27.

82. H. L. Towie, "Motor Menace," *Atlantic Monthly* 136 (July 1925), 98–107; W. L. Chenery, "Motor Massacre," *Collier's* 77 (Mar. 27, 1926), 23; D. Merritt, "Deadly Driver," *Outlook* 142 (Apr. 7, 1926), 521–22; and H. C. Hines, "Morons on the Macadam," *Scribner's* 822 (Nov. 1927), 596–600; D. C. Seitz, "Murder by Motor," *Outlook* 144 (Sept. 22, 1926), 113–14.

83. Richard Shelton Kirby, "The Right to Drive," *Atlantic Monthly* 147 (Apr. 1931), 443.

84. "Where Road Wrecks Happen," *Scientific American* 126, no. 5 (May 1922), 317; see also, Charles F. Barrett, "The 'Blind Turn,' Its Dangers and Various Methods of Solution," *Scientific American* 114, no. 1 (Jan. 1, 1916), 24; and F. D. McHugh, "Horse and Buggy Bridges," *Scientific American* 155, no. 3 (Sept. 1936), 138–41.

85. Bruce E. Seely, *Building the American Highway System: Engineers as Policy Makers* (Philadelphia: Temple University Press, 1987), 72–73.

86. James Wren, "Brakes," in *Encyclopedia of American Business History and Biography: The Automobile Industry, 1896–1920,* ed. George S. May (New York: Bruccoli Clark Layman, 1990), 46.

87. Report on Massachusetts Highway Accident Survey 1934 (Cambridge, MA: Massachusetts Institute of Technology, 1935), 97; see also "State and Municipal Car Inspections Increase Dealer Service Market," *Studebaker Service* 63–A (Sept. 1932), 3, SA.

88. Wren, "Brakes," 46.

89. "Is Your Driving 'Touch' Modern?" *Popular Mechanics* 63, no. 1 (Jan. 1935), 66.

90. Dewey H. Palmer and Laurence E. Crooks, *Millions on Wheels: How to Buy, Drive, and Save Money on Your Automobile* (New York: The Vanguard Press, 1938), iii.

91. See "Danger in Handling Gasoline," in E. V. Burke, *I Know How I Can Find It, I Can Fix It!!* (Stockton, CA: Associated Printing Co., 1930), 115–16.

92. Russell, *Automobile Driving Self-taught,* 64.

93. Ibid.

94. *A Stereoscopic Presentation of Packard Superiorities* (n.p., Packard Motor Car Co., n.d.), unpaginated, "Automobiles," box 10, folder 2, "Packard," WC.

95. Bruce McCalley, *Model T Ford: The Car That Changed the World* (Iola, WI: Krause Publications, 1994), 22–23; "Engine Oil Consumption—1934 Models," *Studebaker Service* (April 1934), 72–A, 4–5, SA.

96. Pagé, *The Model T Ford Car,* 95.

97. Pierce-Arrows for 1913–14, available for prices ranging from $4,850 to $6,000, included a speedometer, "auto-meter" (odometer), clock, and gasoline gauge, but no ammeter. *Pierce-Arrow Model 48–B2, 1913–1914* (Buffalo: The Pierce-Arrow Motor Car Company, n.d.), 11, 31.

98. *The Auto-Meter* (Beloit, WI: Warner Instrument Co., 1910), unpaginated.

99. For early use of "instrument board," see *Automobile* 31, no. 8 (Aug. 20, 1914), 375; and *Paige-Detroit Reference Book, Model "15–19"* (Detroit: Paige-Detroit Motor Car Co., n.d.), 23.

100. At some point in the 1920s, the combination of year-round operation and

a much larger and more competitive used-car market seems to have caused speedometer makers to eliminate the easy reset control on the primary odometer.

101. Sheller, *Me and the Model T,* 191.

102. Edward J. Shipper, "Control Layout Design Analyzed," *The Automobile and Automotive Industries* 37, no. 11 (1917), 435–40.

103. Ray Miller, *Henry's Lady, Model A: An Illustrated History of the Model A Ford* (Oceanside, CA: The Evergreen Press, 1972), 33.

104. Ray Miller, *Chevrolet: The Coming of Age, An Illustrated History of Chevrolet's Passenger Cars, 1911–1942* (Oceanside, CA: The Evergreen Press, 1976), 33–34, 42–43, 64, 99–100, 159, 191, 203–4, 214–15, 223, 236–37, 249.

105. Ibid., 257.

106. Dean A. Fales, "Parts of Vehicles and Their Relation to Accidents," abstracted in Report on Massachusetts Highway Accident Survey 1934 (Cambridge, MA: Massachusetts Institute of Technology, 1934), 81; Newcomb and Spurr, 209.

107. See complaint of Robert F. Chapman in letter to editor, *Automobile* 30, no. 16 (Apr. 16, 1914), 822–23. For illustrations of the "adjustable pedals" on Winton Six and Premier cars, see *Automobile* 26, no. 1 (Jan. 4, 1912), 71.

108. Donald W. Matteson, *The Auto Radio: A Romantic Genealogy* (Jackson, MI: Thornridge Publishing, 1987), 73, 75, 105–11, 115.

109. Herbert H. Evans, *Death and Destruction with Motor Vehicles, How to Prevent* (New Castle, DE: Dale Printing Co., 1931), 8; *Consumers' Research Bulletin* 2, no. 6 (Mar. 1936), 9.

110. *Consumers' Research Bulletin* 2, no. 6 (Mar. 1936), 9, quoting *Consumers' Research Bulletin,* July 1933.

111. Arthur W. Stevens, Highway Safety and Automobile Styling (Boston: The Christopher Publishing House, 1941), 51.

112. John Steinbeck, *East of Eden* (New York: Viking, 1952), 364.

113. Williams v. Russell, 196 Minn. 397, 265 N.W. 270, 273 (1936), quoted in Edward C. Fisher, *Vehicle Traffic Law* (Evanston, IL: The Traffic Institute, Northwestern University, 1961), 10.

Chapter 4. Tools, Tinkering, and Trouble

1. "A Quaker Girl Who Knows Her Car," *Automobile* 58, no. 2 (Jan. 9, 1908), 63.

2. Morris A. Hall, *Care and Operation of Automobiles: A Handbook on Driving, Road Troubles, and Home Repairs* (Chicago: American School of Correspondence, 1912), "Introduction," unpaginated.

3. "Caring for Your Own Car," *Fordowner* (July 1918), 13.

4. Christine Frederick writing in *Suburban Life* (July 1912), 13, quoted in Vir-

ginia Scharff, *Taking the Wheel: Women and the Coming of the Motor Age* (New York: Free Press, 1991), 68.

5. George Harris et al., *Audels Answers on Automobiles for Owners-Operators-Repairmen* (New York: Theodore Audel & Co., 1912), n.p.

6. Recently the phrase has broadened and now refers to the workings or construction of any machine or device. Thus, a 2005 Miele appliance advertisement exhorted buyers, "Before you buy anything . . . check under the hood," while an advocate of open-source computer software derisively asked, "Would you buy a car with the hood welded shut if trying to open the hood could be punished with prison time?" So comfortable are Americans with the metaphor of the automobile hood as the portal beneath which important things are going on that a scholar explained that she sought to "lift the hood, as it were" regarding the interplay between people and their technology, that is, to expose the mechanism behind such interplay and its workings. For the Miele advertisement, see *California Home and Design,* Apr. 2005 (unpaginated advertising section); for the "hood welded shut" software analogy, see Benjamin Mako Hill, "The Geek Shall Inherit the Earth: My Story of Unlearning," note 7, http://yukidoke.org/~mako/writing/unlearningstory/Story OfUnlearning.html, accessed Nov. 13, 2003; and for the scholarly usage, Rosalind Williams, "Historians of Technology in the Information Age," *Technology & Culture* 41, no. 4 (Oct. 2000), 660.

7. "He'd Have to Get Under—Get Out and Get Under (to Fix Up His Automobile)," 1913, box 9, file C, "G–H," DV.

8. Andrew Dyke and George P. Dorris, *Dyke's Diseases of a Gasoline Automobile and How to Cure Them* (St. Louis: A. L. Dyke Automobile Supply Co., 1903).

9. *Whys and Wherefores* (Cleveland: Auto Institute, 1907), unpaginated but item 193 in catechism-like format, JCL.

10. *Fordowner* 7, nos. 11–12 (Aug.–Sept. 1914), 495.

11. H. Clifford Brokaw and Charles A. Starr, *Putnam's Automobile Handbook: The Care and Management of the Modern Motor-Car* (New York: G. P. Putnam's Sons, 1918), 305.

12. Ibid., 57.

13. Hall, *Care and Operation of Automobiles,* 56.

14. *Motor World* 10, no. 3 (Apr. 13, 1905), 113.

15. *Automobile Dealer and Repairer* 19, no. 3 (May 1915), 49.

16. Brokaw and Starr, *Putnam's Automobile Handbook,* iv.

17. Auto Institute, *Whys and Wherefores* (Cleveland, 1907), item 193.

18. "How a Gas Engine Talks," *Popular Mechanics* 7, no. 12 (Dec. 1905), 1144–45; the term *box* may derive from steam engine usage as *crosshead* does.

19. Charles Welsh, ed., *Chauffeur Chaff or Automobilia* (Boston: H. M. Caldwell Co., 1905), 9.

20. H. Clifford Brokaw and Charles A. Starr, *Putnam's Automobile Handbook*, iv; see also, A. L. Brennan Jr., *Automobile Operation* (New York: Outing Publishing Co., 1914), 23–24.

21. See, "They Told Their Troubles," *Motor World* 10, no. 3 (Apr. 13, 1905), 127–28.

22. So-called testing charts, trouble-hunting diagrams, or, in the more martial metaphor preferred after the First World War, troubleshooting guides first appeared in periodicals such as *Horseless Age* and *Scientific American*.

23. *Book of Information Dodge Brothers Motor Vehicles*, 16th ed. (Detroit: Dodge Brothers, Apr. 1923), 121. Similar tool kits came with the Nash; see *Nash Advanced Six, Series 161, Information for the Owner* (Kenosha, WI: The Nash Motor Company, 1924), 65, NYPL.

24. *The Franklin Car, Series 9–B Care and Operation* (Syracuse, NY: Franklin Automobile Company, May 1919), 9, NYPL; and *Franklin Instruction Book, Series Nine* (Syracuse, NY: Franklin Automobile Company, Oct. 1922), 9, WC.

25. "List of Tools" in *Pierce-Arrow Motor Cars, Care and Operation* (Buffalo: Pierce-Arrow Motor Car Co., 1924), 50, FA.

26. *The Packard, no. 62* (Detroit: The Packard Motor Car Company, Sept. 1916), 4.

27. *1911 Catalog of Motor Car Supplies and Machinist Supplies* (Chicago: Motor Car Supply Company, 1911), 180.

28. *Automobile Accessories . . . 1916* (New York: The Motor Car Equipment Co., 1916), 1–4.

29. Winifred Hawkridge Dixon, *Westward Hoboes, Ups and Downs of Frontier Motoring* (New York: Charles Scribner's Sons, 1921), 87.

30. Paul H. Marley, *The Story of an Automobile Trip from Lincoln, Nebraska, to Los Angeles, California, via San Francisco* (n.p., 1912), 11.

31. Vernon McGill, *Diary of a Motor Journey from Chicago to Los Angeles* (Los Angeles: Grafton Publishing Co., 1922), 9.

32. Eric W. Walford, *Maintenance and Repair, Specially Written for the Owner and Driver* (London: Iliffe & Sons Ltd., 1917), 9.

33. Galen Vance Nolan, *Emergency Aids for Stalled Motor Cars, An Up-to-date Selection of Tried and Tested Emergency Aid Helps and Repairs for the Stalled Motor Car* (Lancaster, PA: Merit Publishing Co., 1920), 11.

34. Ibid., p. 23.

35. Vera Marie Teape, "The Road to Denver," 1907, reprinted in the *Palimpsest* (Jan.-Feb. 1980), 10–11, in vertical files, NMAH Transportation Division.

36. Nolan, *Emergency Aids for Stalled Motor Cars*, 11.

37. "Tools for the Home Garage," *Fordowner* (Oct. 1914), 16.

38. Dixon, *Westward Hoboes*, 131.

39. Mort Schultz, "Tires: A Century of Progress," *Popular Mechanics* (June 1985), 63.

40. "Result of Experience, by Owner," *Horseless Age* 21, no. 15 (Apr. 8, 1908), 401–2.

41. Harold W. Slauson, *Everyman's Guide to Motor Efficiency* (New York: Leslie-Judge Co., 1920), 129.

42. For numerous dealer complaints to Ford headquarters about leaky radiators, see accession 94, box 168, 1926 Fishleigh Files, folder: "Complaints, Dealers, 1926," FA; typically, A. T. Smith, a dealer in St. Johns, MI, wrote on Jan. 1, 1926: "Five of the six last radiators installed recently were found to have leaks. There seems to be no particular place. Some of the trouble was in the tubes and some in the solder joints on the tank."

43. "Fuel Pipe Repairs," *Horseless Age* 21, no. 15 (Apr. 8, 1908), 402.

44. Thomas H. Russell, *Automobile Driving Self-taught: An Exhaustive Treatise on the Operation, Management, and Care of Motor Cars* (Chicago: Charles C. Thompson Co., 1909), 51; see also, Slauson, *Everyman's Guide to Motor Efficiency,* 136, and Harold P. Manly, *The Modern Motor Car: A Book of Simplified Upkeep* (Chicago: Laird & Lee, Inc., 1914), 19.

45. Victor Pagé, *Automobile Repairing Made Easy* (New York: Norman W. Henley Publishing Co., 1919), 354; for utility of chewing gum and bread paste, see Thomas H. Russell, *Questions and Answers for Automobile Students and Mechanics* (Chicago: Charles C. Thompson Co., 1911), 86; for folk polishing remedy, "100 Tips for Readers of Fordowner," *Fordowner* 6, no. 1 (Oct. 1916), 35–36.

46. Dr. D. Birkhoff, "Six Years' Experience with the Automobile," *Horseless Age* 21, no. 1 (Jan. 1, 1908), 5–6.

47. *Nash Advanced Six, Series 161 Information for the Owner* (Kenosha, WI: Nash Motor Corporation, 1924), 66.

48. Letter to editor from F. X. K., *Motor* 24, no. 21 (May 25, 1911), 1187.

49. Brokaw and Starr, *Putnam's Automobile Handbook,* 268.

50. Murray Fahnestock, "The Search for Silence," *Fordowner* 7, no. 2 (May 1917), 52, 54, 56, 58, 60, 62, 64; *Fordowner* 3, no. 6 (Sept. 1915), 30.

51. Paul E. Vernon, *Coast to Coast by Motor* (New York: William Edwin Rudge, 1930), 101.

52. *Buick Reference Book, Models D-4-34 and D-4-35* (Flint, MI: Buick Motor Co., 1916), 47.

53. J. C. Kohr of Kansas City to Ford, Dec. 12, 1928, accession 97, box 207, "Foster Files, Brakes-Complaints," folder: "Complaints," FA.

54. Letter from Charles H. Bohn to Ford Motor Company, July 13, 1928, accession 94, box 207, folder: "Criticisms, General," FA.

55. Murray Fahnestock, "Causes and Cures of Overheating," *Fordowner* 7, no. 4 (July 1917), 27.

56. New York Automobile Assn., *Motordom* 7, no. 5 (Sept. 1913), 110.

57. See David N. Lucsko, *The Business of Speed: The Hot Rod Industry in America, 1915–1990* (Baltimore: Johns Hopkins University Press, 2008), 1–39.

58. Frank N. Blake, "Some Road Experiences in 1903," *Horseless Age* 13, no. 8 (Feb. 24, 1904), 218.

59. J.E.B. to editor, *Automobile* 29, no. 23 (Dec. 4, 1913), 1062; *Helix Gas Mixer* (n.p., n.d.), unpaginated brochure "Automobile Industry," box 28, WC; Owner experiments with kerosene as a coolant created more problems than they solved. Although the liquid did not easily freeze, it boiled at a much lower temperature than water and thereby was less effective in drawing heat from the engine. Even worse kerosene dissolved the cooling system's rubber hoses, turning them to jelly.

60. Morris A. Hall, "Accessories of Value and Advice for Installing Them on Automobiles," *Motor Car Overhauling* (Pawtucket, RI: Automobile Journal Publishing Co., n.d.), 42–43.

61. *Automobile* 29, no. 9 (Aug. 28, 1913), 388.

62. *Fordowner* 9, no. 6 (Sept. 1918), 111.

63. "100 Tips for Readers of Fordowner," *Fordowner* 6, no.1 (Oct. 1916), 35.

64. Pagé, *Automobile Repairing Made Easy*, 209.

65. "Carbon Deposits," *Fordowner* 3, no. 6 (Sept. 1915), 47.

66. E. W. Jones to *Horseless Age* 24, no. 3 (July 21, 1909), 78.

67. National Illustrating Co., *Automobile Emergency and Repair Manual: A Manual of Information to Meet Every Conceivable Emergency* (New York: 1911), 17.

68. Slauson, *Everyman's Guide to Motor Efficiency*, 41.

69. C. R. Strouse, *Automobile Operation and Repair* (Scranton, PA: International Textbook Company, 1924), section 13, 16.

70. "Greatest Trouble Causer Is Carbon," *Fordowner* 9, no. 4 (Aug. 1918), 80.

71. *Fordowner* 5, no. 1 (Apr. 1916), inside back cover.

72. *Fordowner* 1, no. 1 (Apr. 1914), 24.

73. "They Told Their Troubles," *Motor World* 10, no. 3 (Apr. 13, 1905), 127–28.

74. Letter from William J. Nutty to *Automobile* 25, no. 21 (1911), 822.

75. *Instructions Concerning the Care and Operation of Franklin Motor Cars, All Models, 1909* (Syracuse, NY: H. H. Franklin Manufacturing Co., 1909), unpaginated; Robert E. Sherwood, *The Care of Automobiles, with a List of Don'ts for Motor Car Drivers and Acute Automobilia* (New York: Neis Publishing Co., 1911), 4.

76. Herbert L. Towle, "The Carbureter, Its Adjustment, and the Novice," *Automobile* 22, no. 13 (Mar. 31, 1910), 630–31.

77. *Cadillac Operators' Manual Type 61* (Detroit: Cadillac Motor Car Co., 1923), 6; *Instruction Book for Locomobile (Six-Cylinder) Cars* (Bridgeport, CT: The Loco-

mobile Company of America, 1915), 59, WC; *Operation and Care National Sextet* (Indianapolis: National Motor Car & Vehicle Corp., c. 1921), 4.

78. *Zenith Carburetor (Bavery's System) Instructions for Operating and Adjusting,* 7th rev. ed. (Detroit: Zenith Carburetor Co., 1919), 5.

79. A. S. Logan to *Automobile* 18, no. 2 (Jan. 9, 1908), 46.

80. Raymond H. Maulsby, *Automotive Troubles: The Professional "Trouble Shooter"* (Kansas City, 1924), 7.

81. Quoted in Ralph D. Gray, *Alloys and Automobiles: The Life of Elwood Haynes* (Indianapolis: Indiana Historical Society, 1979), 108.

82. *Horseless Age* 25, no. 3 (Jan. 19, 1910), 39.

83. *Owners' Instruction Book for the Operation and Care of the Flint Model "80"* (Flint, MI: Flint Motor Car Co., n.d.), 49, WC.

84. *Studebaker Service* (South Bend, IN: Studebaker Motor Car Co.), 49–A (Oct. 1930), 14, SA.

85. This development is nicely documented in the small Illinois town of Oregon, about one hundred miles from Chicago; see Norman T. Moline, *Mobility and the Small Town, 1900–1930,* Research Paper 132 (Chicago: University of Chicago Department of Geography, 1971), Table 13, 156–57.

86. "Kind of Highway Determines Type of Successful Station," *National Petroleum News* 20 (June 6, 1928), 29–30, quoted in John A. Jakle and Keith A. Sculle, *The Gas Station in America* (Baltimore: Johns Hopkins University Press, 1994); see also Table 3.1, 58.

87. Steve McIntrye, " 'The Repair Man Will Gyp You': Mechanics, Managers, and Customers in the Automobile Repair Industry, 1896–1940" (Ph.D. thesis, University of Missouri-Columbia, 1995), 60, 127, 240.

88. Sigmund Krausz, *Krausz's ABC of Motoring: A Manual of Practical Information for Layman, Auto Novice and Motorist* (Chicago: Laird & Lee, 1906), 105; Writing in the 1970s, William J. Abernathy sees the American automobile as having reached a "plateau" of development by the time of the war: "Technological Change in the U.S. Automobile Industry: A Historical Overview," in his *Productivity Dilemma: Roadblock to Innovation in the Automobile Industry* (Baltimore: Johns Hopkins University Press, 1978), 20.

89. Henry C. Pearson, *Pneumatic Tires: An Encyclopedia* (New York: The India Rubber Publishing Co., 1922), 47; Dixon, *Westward Hoboes,* 66; Rudi Volti, *Cars and Culture: The Life Story of a Technology* (Westport, CT: Greenwood, 2004), 10.

90. Apr. 8, 1924, 1, accession 78, box 74, "Service, General Letters," FA.

91. That a 1952 compilation of motorist advice published by *Popular Mechanics* contained not a single reference to squeaks, rattles, or other body and chassis noise attests to the success of the search for silence.

92. Mary Crehore Bedell, *Modern Gypsies: The Story of a Twelve-Thousand-Mile Motor Camping Trip Encircling the United States* (New York: Brentano's, 1924), 140.

93. *Automotive Engineering,* vol. 1 (Chicago: American Technical Society, 1927), 291.

94. Robert W. A. Brewer, "The Utilization of Liquid Fuel," *Antique Automobile* 18, no. 2 (Summer 1954), 61–62.

95. Dixon, *Westward Hoboes,* 2.

96. *Antique Automobile* 9 (Nov. 1945), 2, 4.

Chapter 5. Reading the Owner's Manual

1. *Motor Age* 2, no. 21 (Nov. 20, 1902).

2. For a single, broadside-sheet-sized set of instructions printed on fabric, see *Directions for Using Colt's Pistols, Rifles, Carbines, and Shot Guns* (Hartford, CT: Colt's Armory, n.d. but after 1861), in Division of Military History, NMAH (shown to me by David Miller); and the three-page brochure, *Description and Rules for the Management of the Springfield Breech-Loading Rifle Musket, Model 1866* (Springfield, MA: United States Armory, 1867; reprint edition Pioneer Press, Harriman, TN, n.d.); *Studebaker Service* 18A (June 1927), 2.

3. Advertising flyer, "Dr. A.L. Dyke (Motor Doctor)," n.d., in Dyke Collection folder: "Office Specialty Manufacturing Co. New Market Ont. Canada," DPL.

4. Charles Roswell Henry to Winton Motor Company, Nov. 18, 1905, Henry Papers, letterpress book, BL.

5. S. J. Fort, M.D., "Why the Second-Hand Car?" *Fordowner* 8, no. 2 (May 1918), 31–32; for an example of the catechism approach, see Victor W. Pagé, *Questions and Answers Relating to Modern Automobile Design, Construction, Driving, and Repair* (New York: Norman W. Henley Publishing Co., 1919).

6. "More and Better Instruction Books," *Motor* 2, no. 21 (Nov. 20, 1902), 4.

7. "Introduction," *Whys and Wherefores of the Automobile, A Simple Explanation of the Elements of the Gasoline Motor Car, Prepared for the Non-Technical Reader* (Cleveland: Auto Institute, 1907), unpaginated, JCL.

8. A. Hyatt Verrill, *A-B-C of Automobile Driving, A-B-C Series* (New York: Harper & Bros., 1916); M. E. Smoyer, *Auto Repairing Simplified* (Kansas City: Tiernan-Dart Printing Co., 1918); Harold Whiting Slauson, *First Aid to the Car, or, Highway Hints and Helps* (New York and London: Harper & Brothers Publishers, 1921); Harold W. Slauson, *Everyman's Guide to Motor Efficiency: Simplified Short-Cuts to Maximum Mileage at Minimum Cost* (New York: Leslie-Judge Co., 1920).

9. McGraw-Hill *Book Notes* (New York: McGraw-Hill, Jan. 1923), 11.

10. Victor W. Pagé, *The Model T Ford Car* (New York: Norman W. Henley Publishing Co., 1926), 11–12.

11. Joseph Tracy, "Common Sense in Automobile Driving" (Nov. 1907), 35–38; 102; C. H. Claudy, "Learning to Drive a Motor Car" (Sept. 1908), 469–70.

12. Julian Street, "Good and Bad Driving: The Pneumatic Peril of the Highway," *Collier's* (Jan. 7, 1911), Supp., 9; Charles P. Klein, "The Mistakes of Beginners" *Literary Digest* (Aug. 10, 1912), 230, 232, 234. "Little Things about a Car That Every Woman Who Means to Drive One Ought to Know," *Ladies' Home Journal* (Mar. 17, 1917), 32.

13. Warren I. Sussman, *Culture as History: The Transformation of American Society in the Twentieth Century* (New York: Pantheon, 1984), 106–7; Susan Strasser, *Satisfaction Guaranteed: The Making of the American Mass Market* (New York: Pantheon, 1989), 131; Burton J. Bledstein, "Introduction: Storytellers to the Middle Class," in Burton J. Bledstein and Robert D. Johnston, eds., *The Middling Sorts: Explorations in the History of the American Middle Class* (New York: Routledge, 2001), 10.

14. Charles Bazerman, *Shaping Written Knowledge: The Genre and Activity of the Scientific Article in Science* (Madison: University of Wisconsin Press, 1988), 11.

15. *1914 Cadillac Instruction Book* (Clymer reprint), 8; *Instructions for Care and Operation Cadillac 1914*, 3rd ed. (Detroit: Cadillac Motor Car Co., 1914), 72. For a similar statement, see *A Study of the Oldsmobile: A Little Booklet for Users of the Oldsmobile Curved Dash Runabout* (Detroit: Olds Motor Works, n.d.), title page, FA.

16. Ibid., front cover.

17. *The Franklin Car, Series 9–B, Care and Operation* (Syracuse, NY: Franklin Automobile Company, Sept. 1920), 7; the 1915 Locomobile came with an "instruction book on ignition" as well as a regular manual.

18. Beverly Rae Kimes and Henry Austin Clark Jr., *Standard Catalog of American Cars, 1805–1942* (Iola, WI: Krause Publications, 1985), 985.

19. *Study of the Oldsmobile*, 5.

20. John R. Brockmann, *From Millwrights to Shipwrights to the Twenty-first Century: Explorations in a History of Technical Communication in the United States.* Written Language Series, edited by Marcia Farr (Cresskill, N.J.: Hampton Press, Inc., 1998), chap. 5.

21. Of the hundreds of owner's manuals consulted for this book only one, *Manual of Instruction on Krit Motor Cars*, identified an author by name, saying on its first page that it was "written for the Krit Motor Car Co. by Lee A. Cuson," *A Manual of Instruction on Krit Motor Cars, All Models* (Detroit: Krit Motor Car Co., June 1913), front page, PCA; the one pre-1940 manual I found where the department within the automaker's organization was identified was the *Instruction Book for Locomobile (Six-Cylinder Cars, The 38L, Type R6, The 48 L, Type M6* (Bridgeport, CT: Locomobile Corporation of America, May 1915), 4, NMAH, where the title page states that it was published by the "advertising department."

22. Whereas I see automobile owner's manuals in the years before about 1910 as significantly inferior to those for sewing machines, Brockmann only looked at auto owner's manuals after 1912, a focus that contributed to his emphasis on the continuity of effective communication between the manuals of the sewing machine industry and those of the auto industry. Moreover, by looking only at the manuals published by Ford and Chevrolet, he missed the greater variation within the industry.

23. Not surprisingly, given the vast numbers of automobile manufacturers that came and went in these years, some later manuals still were exceptionally thin and poor. That for the 1917–18 Nelson had but twenty-three pages and no illustrations, see: *Nelson Instruction Book, Operation and Care of the NELSON "Four-29" 1917–18* (Detroit: E. A. Nelson Motor Car Company, 1917).

24. *Instruction Book Applying to the Model 47* (Lansing, MI: Oldsmobile Motor Works, July 1921); and *Oakland Instruction Book and Repair Parts List, Model 6–44* (Pontiac, MI: Oakland Motor Car Company, n.d.), NMAH; Brockmann, again confining his examination to Ford and Chevrolet, determined the average manual as increasing to fifty or sixty pages by the 1920s, Brockmann, *Technical Communication*, 232.

25. *Ford Model T Instruction Book,* 6th ed. (Detroit: Ford Motor Company, May 1913; reprint), 16–17.

26. Ibid., 15, 22, 33, 41, 44.

27. Ibid., 8.

28. *Instruction Book Applying to the Oldsmobile 'Six' Model 37* (Lansing, MI: Oldsmobile Motor Works, 1917).

29. *Studebaker—Models EK and EL 24 Series* (South Bend, IN: Studebaker Corporation, 1922–24).

30. As early as 1908 this sort of deductive thinking was itself codified in a visual form, in what were usually called "trouble-hunting diagrams." See, "A Simplified Method of 'Trouble Hunting,'" *Horseless Age* 21, no. 6 (Feb. 5, 1908), 152–53, reproducing a trouble-hunting diagram that originally appeared in a French publication, *Omnia,* which had offered a prize to readers for "the best system of trouble hunting." For an early example of such a diagram in an owner's manual, see *A Manual of Instruction on Krit Motor Cars, All Models* (Detroit: Krit Motor Car Co., 1913), 72–73, PCA.

31. *Book of Instruction, Lincoln Motor Cars,* 1st ed. (Detroit: Lincoln Motor Company, Jan. 1, 1922), 38–39; *Book of Instruction, Lincoln Motor Cars,* 5th ed. (Detroit: Lincoln Motor Company, Jan. 1, 1925), 38–39; *Book of Instruction, Lincoln Motor Cars,* 6th ed. (Detroit: Lincoln Motor Company, May 1, 1927), 35, NMAH; *Studebaker Service,* 18–A, June 1927, 2, SA.

32. *Ford Model "A" Instruction Book* (Detroit: Ford Motor Co., n.d.), thanks to

Peter Jakab for loaning me this manual, which is original to his unrestored 1930 Ford Model A Tudor; *Instruction Book Hudson Super-Six, 1928 Model 'O' and Model 'S,' September 1928* (Detroit: Hudson Motor Car Company, 1928), "foreword," unpaginated.

33. *Book of Instructions Chalmers "Six-30"* (Detroit: Chalmers Motor Car Company, n.d. but c. 1916); *Book of Instruction Lincoln Motor Cars,* 3rd ed. (Detroit: Lincoln Motor Company, May 1, 1922), NMAH.

34. *Instruction Book Hudson Super-Six* (Detroit: Hudson Motor Car Co., 1928); *Instruction Book Jordan Model J Starting with Serial Number 70,001* (Cleveland: Jordan Motor Co., Inc., n.d.); *Instruction Book for Locomobile Six-Cylinder Cars* (Bridgeport, CT: The Locomobile Company, 1915); *Instruction Book for the Guidance of Operators of the 1907 Models Peerless Cars* (Cleveland: The Peerless Motor Car Company, 1907); *Matheson Instruction Book* (Wilkes-Barre, PA: Matheson Motor Car Co., 1907), NMAH.

35. *Book of Information Dodge Brothers Motor Vehicles,* 14th ed. (Detroit: Dodge Brothers, Oct. 1, 1921), WC.

36. *Paige-Detroit Reference Book Model "16–19"* (Detroit: Paige-Detroit Motor Car Company, n.d.), NMAH.

37. *Instructions for Series 10 Franklin Cars* (Syracuse, NY: Franklin Automobile Co., n.d.); *Instructions for the Care and Operation of the Stanley Steam Car without Condenser* (n.p., 1913; Floyd Clymer reprint), NMAH.

38. *Nash Advanced Six Series 161 Information for the Owner* (Kenosha, WI: Nash Motor Company, 1924); *1938 Buick Owner's Manual* (Flint, WI: Buick Motor Division, 3rd. edit., 1937); *Hudson 112 Owner's Manual, 1938* (Detroit: Hudson Motor Car Co., 1937); The *1938 Buick Owner's Manual* contained a "special section" for what it called "the Mechanically Minded," relegated to the "back part of the book," or pages 77–92.

39. *Your New 1942 Nash* (Kenosha, WI: Nash Motors, 1941), 3.

40. Harry A. Tarantous, "Driving an Automobile" *Good Housekeeping* 64 (June 1918), 128–30.

41. Ibid.

42. John Chapman Hilder, "Keep Your Car from Catching Cold," *Ladies' Home Journal* 37 (Dec. 1920), 44, 186.

43. Myron M. Stearns, "Learning to Drive," *Saturday Evening Post* 198 (June 19, 1926), 213.

44. See also, Tarantous, "Driving an Automobile," 129–30.

45. Will Irwin, "Mountain Motoring," *Saturday Evening Post* 195 (Nov. 11, 1922), 37.

46. Courtney Ryley Cooper, "Slow—Sound Horn," *Saturday Evening Post* 194,

no. 46 (May 13, 1922), 19; for other examples of advice on mountain driving, see Stearns, "Learning to Drive," 213; Tarantous, "Driving an Automobile," 129.

47. Cooper, "Slow—Sound Horn," 104.

48. Ibid., 106.

49. Martin Bunn, "Is Your Car in Summer Trim"? *Popular Science* 107, no. 1 (July 1925), 73.

50. Martin Bunn, "Gus and Joe Are Real Live Men," *Popular Science* 117, no. 2 (Aug. 1930), 70.

51. "Gus Tells How to Adjust a Carburetor and Shows How to Cure Starting Trouble," *Popular Science* 108, no. 4 (Apr. 1926), 66, 137–38.

52. *Popular Science* 108, no. 1 (Jan. 1926), 66, 145–46, 109; no. 5 (Nov. 1926), 62, 164. *Popular Science* 111, no. 1 (July 1927), 69–70; no. 3 (Sept. 1927), 69–70.

53. "Gus Turns High-Pressure Salesman [about the thieving gypsy, 'Tony']," *Popular Science* 147, no. 1 (July 1945), 138–40; "How to Save Money on Tires: Under Inflation and Neglect Cost You Miles and Dollars [featuring the 'penny-wise and pound-foolish' Mr. Burr]," *Popular Science* 110, no. 6 (June, 1927), 71–72; "Gus Wilson and the DeePee [involving a displaced person, a foreign painter]," *Popular Science* 152, no. 2 (Feb. 1948), 172–75; "Your Car Is as Old as You Make It [involving a stuck-up, know-it-all physician]," *Popular Science* 110, no. 3 (Mar. 1927), 65, 154, 156; "Don't Be a Back-Seat Driver! Gus Helps a Husband Who Was Nagged into a Smash and Gives Some Advice for Passengers," *Popular Science* 111, no. 4 (Oct. 1927), 69, 182.

54. "Gus Tunes a Car by Ear," *Popular Science* 133, no. 3 (Sept. 1938), 54, 97–98.

55. *Popular Science* 128, no. 2 (Feb. 1936), 56, 120–21; and 129, no. 1 (July 1936), 54, 114.

56. *Popular Science* 129, no. 2 (Aug. 1936), 54, 108; no. 4 (Oct. 1936), 60, 129–30; and no. 5 (Nov. 1936), 60, 131–32.

57. See, for example, "How to Pick the Best Car in Its Class," *Popular Science* 108, no. 5 (May 1926), 44–45; and "Can You Afford a New Car?" *Popular Science* 129, no. 2 (Aug. 1936), 54, 108.

58. "Heat Your Car but Don't Burn It!" *Popular Science* 118, no. 2 (Feb. 1931), 86.

59. Gregory L. Williams, "Consumer's Research History," An Inventory to the Records of Consumers' Research, Inc., 1910–1983, bulk 1928–1980 (Rutgers University), http://www2.scc.rutgers.edu/ead/manuscripts/consumers_introf.html, accessed Sept. 15, 2008.

60. "Automobiles of 1936," *Consumers' Research Bulletin*, n.s., 11, no. 6 (Mar. 1936), 3–22.

61. Charles F. McGovern, "Sold American: Inventing the Consumer, 1890–1940" (Ph.D. thesis, Harvard University, 1993), 307, 309.

62. "MI Tests the New Cars, Ford and Buick," *Mechanix Illustrated* 35, no. 2 (Feb. 1946), 48–51, 74.

63. Ibid., 51.

64. "MI Tests the New Cars, Oldsmobile, Packard," *Mechanix Illustrated* 35, no. 4 (Apr. 1946), 52–53, 155.

65. Ibid., 54–55, 156.

66. "MI Tests the '48 Hudson," *Mechanix Illustrated* 39, no. 5 (Mar. 1948), 80–81, 92.

67. Ibid., 81.

68. Floyd Clymer wrote *Popular Mechanics'* first "owners reports," as the magazine called its road tests, in 1951. *Road and Track* was one of the first postwar automotive periodicals, testing the 1946–47 Ford in its premier issue in June 1947 (11–12); other new postwar automobile magazines that regularly published road tests included *Motor Trend* (1949) and *Car and Driver* (1955).

69. David E. Nye, *Image Worlds: Corporate Identities at General Electric* (Cambridge, MA: MIT Press, 1985), 27.

70. "Camera Road Test: Yashica T4, Klaxons' Quiet One Muses on His Little-Known Secret Passion," *PsychoPEDIA*, http://www.psychopedia.com/dailynews/2007/11/camera_roadtest_yashica_t4the.html, accessed Apr. 21, 2009.

71. For Apple ad, see "New Software Assists Customer's Test Drive," *Apple-Gram*, Nov. 15, 1984; for PCBrands example, see *PC/Computing* 5, no. 7 (July 1992), 185.

72. Peter H. Lewis, "New PC: How to Kick the Tires," *New York Times*, Dec. 3, 1998, D1, 1; Jeff Hensen, marketing manager for Gateway Computer, quoted in Michel Marriott, "Thinking Outside the Beige Box," *New York Times*, June 17, 1999, "Circuits," 1.

Chapter 6. Computers and the Tyranny of Technology Consumption

1. Crickett Townsend, "President's Message," *SPC Apples* (Fremont, CA), 3 (Dec. 1984), 3, series 12, box 45, folder: "SPC Apple Vol 3 No 1–4, 6, 10 1984," ACR.

2. As early as 1961, according to Lizabeth Cohen, citing the Kerner Commission's post riot report, 95 percent of poor New Yorkers had television sets, "ensuring that few blacks were deprived of images of American abundance"; quoted in her *A Consumers' Republic: The Politics of Mass Consumption in Postwar America* (New York: Knopf, 2003), 373.

3. "Servants of Your Light Socket," *Radio Broadcast* 11, no. 6 (Oct. 1927), 356.

4. Giles Slade, *Made to Break: Technology and Obsolescence in America* (Cambridge, MA: Harvard University Press, 2006), 106.

5. Paul E. Ceruzzi, *A History of Modern Computing* (Cambridge, MA: MIT Press, 1999), 15.

6. Many have told the story of the personal computer, coming at it from various angles. Two recent and important studies are Fred Turner, *From Counterculture to Cyberculture: Stewart Brand, the Whole Earth Network, and the Rise of Digital Utopianism* (Chicago: University of Chicago Press, 2006); and Thiery Bardini, *Bootstrapping: Douglas Engelbart, Coevolution, and the Origins of Personal Computing* (Palo Alto: Stanford University Press, 2000); see also Bryan Pfaffenberger, "The Social Meaning of the Personal Computer: Or, Why the Personal Computer Revolution Was No Revolution," *Anthropological Quarterly* 61, no. 2 (Jan. 1988), 39–47; and Ceruzzi, *Modern Computing,* especially chap. 7.

7. Ceruzzi, *Modern Computing,* 229.

8. Thanks to Paul Ceruzzi, who personally related by e-mail the story of how people in the Clemson University agriculture program programmed an Altair to control the lights, heating, and ventilation in their greenhouse.

9. Martin Campbell Kelly and William Aspray, *Computer: A History of the Information Machine* (New York: Basic Books, 1996), 237.

10. Frank Rose, *West of Eden: The End of Innocence at Apple Computer* (1989), 62, quoted in Steve Weyhrich, *Apple II History,* www.blinkenlights.com/classiccmp/apple2history.html (Zonker Software, 1991), 6.

11. Stan Veit, *Stan Veit's History of the Personal Computer: From Altair to IBM, A History of the PC Revolution* (Alexander, NC: WorldComm, 1993), 99.

12. In *Made to Break: Technology and Obsolescence in America* (Cambridge, MA: Harvard University Press, 2006), Giles Slade claims that the "killer app that would drive computers into virtually every small business office in America, and eventually into American homes" was word processing packages such as WordStar, 207.

13. "Industry Overview," 24, series 7, box 14, folder: "Fact Book Personal Computer Market [1984?] 1 of 3," ACR.

14. "Copy Strategy, Feb. 14, 1983," 1, series 7, box 14, folder: "Fact Book Personal Computer Market [1984?] 3 of 3," ACR.

15. "Industry Overview," 6, series 7, box 14, folder: "Fact Book Personal Computer Market [1984?] 1 of 3," ACR.

16. Regarding the census data, see http://maisonbisson.com/blog/post/11088/us-census-on-internet-access-and-computing/, accessed July 12, 2010.

17. Otto Friedrich, "The Computer Moves In," *Time,* Jan. 3, 1983, http://www.time.com/time/magazine/article/0,9171,952176–2,00.html, accessed Dec. 9, 2008; for anecdote about Friedrich and typewriters, see the memorial comment following his death, "To Our Readers, May 8, 1995," http://www.time.com/time/magazine/article/0,9171,982897,00.html, accessed Aug. 20, 2010.

18. "Industry Overview," 18, series 7, box 14, folder: "Fact Book Personal Computer Market [1984?] 1 of 3," ACR.

19. "Tandy Takes on the World: The Tenth Anniversary Announcements," 2, 9, series 7, box 4, "Marketing and Sales Newsletters," folder: "Market Intelligence ROM, June–December 1987," ACR.

20. Letter from "An Apple Friend," *Applesource* 6 (Sept. 1979), 2, series 7, box 2, folder: "Apple Source: A Semi-Regular Newsletter for Apple Computer Dealers, 1 of 3," ACR.

21. "Coping with the Kids," *Applesource* 16 (Feb. 1981), 2, series 7, box 2, folder: "Apple Source: A Semi-Regular Newsletter for Apple Computer Dealers, 1 of 3," ACR.

22. Erik Sandberg-Diment, "The Children Seem to Know No Fear," *New York Times* (Aug. 17, 1982).

23. Jeanne Boucher, "From One Green Apple to Another," *Apple Barrel* (Houston-area Apple Users Group), 8, no. 7 (July 1985), 6, series 12, box 4, folder: "Apple Barrel vol. 8 No 6–7, 9 1985 June–Nov," ACR.

24. Erik Sandberg-Diment, "How Much Memory to Buy," *New York Times* (June 22, 1982).

25. Cartoon by Modell, *The New Yorker*, 1980, reproduced in *Applesource* 17 (Mar. 1981), 4, ACR.

26. For information on Stanford's effort to nudge humanists into computing, see Norman K. Wessells, Dean of the School of Humanities and Sciences, Stanford University, "Memorandum to Faculty Members in the Humanities, Oct. 19, 1982," copy in author's possession. For background on the "Tiro Project," as the university's effort was called, and an analysis of how professors conceived of computers and worked with them, see also the report commissioned by the university and written by Peter Lyman, "The Tiro Project: The Planning and Training Phases," unpublished manuscript, July 7, 1983, copy in author's possession.

27. Erik Sandberg-Diment, "A Bundle of Disks for Those Bewildered Novices," *New York Times*, Aug. 8, 1983.

28. The second computer, which went into her study, was an IBM-XT, the first model from the company having a hard drive rather than one or two floppy disk drives.

29. "IBM PC AT Competitive Evaluation," 13, series 7, box 2, folder: "Competitive Evaluation Read Only Memo News Brief 1983, 1 of 4," ACR.

30. This message was replaced by an even scarier version, "Abort, Retry, Fail?" with the release of MS-DOS 3.30 in 1987.

31. Robert J. Sawyer, "WordStar: A Writer's Word Processor," 1990, 1995, at http://www.sfwriter.com/wordstar.htm, accessed Nov. 13, 2008.

32. Victor Frank, "Wordstar Command Summary," reprinted from Sanyo Hack-

er's Newsletter 13, no. 8 (Dec. 1996), http://www.glinx.com/~grifwood/WSCOM SUM.PDF, accessed Dec. 26, 2008.

33. "ESD: Electrostatic Discharge—It Kills, It Maims, It Corrupts," *Adam and Eve* 10, no. 2 (Feb. 1988), 6, series 12, box 3, folder 8, "Adam and Eve, Apple Users' Group vol. 5, no. 4–6, 8, 10, 11," ACR.

34. Steve George, "That Time Again," *Minni'app'les* 7, no. 1 (Jan. 1984), 23, series 12, box 33, folder: "Mini'app'les Vol 7, No 1–5 1984 Jan–May," ACR.

35. "Have you checked your floppies lately?" *Mad Mac News* 4, no.5 (May 1988), 21, series 12, box 33, folder: "Mini'app'les Vol 7, No 1–5 1984 Jan–May," ACR.

36. Software manufacturers soon began encrypting, or locking, their programs to prevent such copying. Some sought to prevent buyers from copying programs and giving them away to friends or even selling them. Others, taking a harsher position, believed as did a vice president of "one well-known manufacturer" who posed the question, "If you buy a Chevrolet and crack it up, you don't expect General Motors to supply you with a new car. Why do you expect me to supply you with a new copy of my program if you spill coffee on the disk?" Barry D. Bayer, "It Seems to Me . . ." *Mini'app'les* 6, no. 6 (June 1983), 26, 28, series 12, box 33, folder 6, "Mini'app'les, Vol 6, No. 1, 5–6 1983 Jan–June," ACR.

37. Veit, *Veit's History of the Personal Computer,* 289–91.

38. Tom Weishaar, "Solving Printer Problems," reprinted from Open-Apple (Overland Park, KN) in *Apple Barrel* 8, no. 9 (Oct.-Nov. 1985), series 12, box 4, folder 11, 25, ACR.

39. "Service in the '80s," *Applesource* 9 (Feb. 1980), 4, series 7, box 2, folder: "Apple Source: A Semi-Regular Newsletter for Apple Computer Dealers, 1 of 3," ACR.

40. Erik Sandberg-Diment, "A Bundle of Disks for Those Bewildered Novices," *New York Times,* Aug. 23, 1983.

41. Ellen Rose, *User Error: Resisting Computer Culture* (Toronto: Between the Lines, 2003), 112. Rose says the term *user-friendly* was first "coined by programmers as a disparaging reference to user ineptitude" but later was co-opted by advertisers and marketing men in their effort to sell computer products to skeptical neophytes.

42. Weyhrich, *Apple II History,* 6.

43. Erik Sandberg-Diment, "The Helpless Feeling," *New York Times,* Sept. 28, 1982.

44. Walter A. Ettlin, *WordStar Made Easy* (New York: McGraw-Hill Osborn Media, 1982); Miranda Morse, *WordStar in Three Days: What to Do When Things Go Wrong* (Huntington, NY: Maple Hill Press, 1984); Don Cassell, *WordStar Simplified for the IBM Personal Computer* (Englewood Cliffs, NJ: Prentice-Hall, 1984).

45. Dick Baumann, "Service Updates," *Applesource* 12 (n.d. [1980]), 7, series 7,

box 2, folder: "Apple Source: A Semi-Regular Newsletter for Apple Computer Dealers, 1 of 3," ACR.

46. "The Apple Hot Line Needs Your Help!" 7, *Applesource* 12 (n.d. [1980]), 7, series 7, box 2, folder: "Apple Source: A Semi-Regular Newsletter for Apple Computer Dealers, 1 of 3," ACR.

47. Service Center, New Help from Apple Service and Support," *Applesource* 3 (Jan. 1983), "7, series 7, box 2, folder: "Applesource 1983–1985, 3 of 3," ACR.

48. Jean Mickelson, "MacTips," *Mad Mac News* [Madison, WI] 3, no. 3 (March 1988), 8, series 12, box 32, folder 2, "Mad Mac News Vol 2 No 1–3, 5, 10–12 1987 Jan–Dec," ACR.

49. The original Macintosh did not have a hard drive like the IBM. It stored its operating system on a removable internal disk, but all other programs and data had to be stored on floppies. In 1987 Apple brought out its Macintosh SE, the first of its products to offer a hard drive (20 or 40 MB) or, if preferred, two floppy drives. Wikipedia, http://en.wikipedia.org/wiki/Macintosh_SE, accessed Jan. 3, 2009.

50. For the argument that DOS provided users with greater control and transparency, see Neal Stephenson, *In the Beginning Was the Command Line* (New York: Avon, 1999).

51. Roy A. Allan, *A History of the Personal Computer: The People and the Technology* (London, Ont.: Allan Publishing, 2001), 12, 18.

52. Human or "user error" allegedly causes 32 percent of data loss, and from personal experience I know how easy it is to make such errors. "Data Loss Statistics," July 9, 2008, http://www.datadepositbox.com/Index.php/component/option,com_lmojo/itemid,47/p.20/, accessed Mar. 25, 2009.

53. Ibid. Software such as Apple's "Time Machine," which seamlessly backs up data automatically to an external hard drive, represents another user-friendly advance. Such online storage to a remote server, or "in the cloud," is an even greater boon to users because not only is it automatic but it also guards against data loss in the worst-case scenario of losing everything—one's computer, hard drive, external storage devices, and removable media—in a fire or flood.

54. Reid Goldsborough, "Computer Problems Can Infuriate the Most Tech Savvy," *Black Issues in Higher Education* 21, no. 1 (Feb. 26, 2004), 37.

Epilogue. The Technology Treadmill

1. Honeywell first introduced the design, which, its corporate website claims, "adorns the walls of more households around the world than any other thermostat," in 1953; http://www51.honeywell.com/honeywell/about-us/our-history.html, accessed July 20, 2010; thanks to Carlene Stephens for this reference.

2. Gruen adapted the old technology to electronic and digital conditions in its

1974 LCD watch. The timekeeper had a stem that when turned a quarter revolution to the right caused the hour to advance once a second; when pulled out to "Position B" and turned, the stem caused the minute to increase once per second. Pushing and turning the stem back to its original starting point returned the watch to its default, timekeeping mode. See *LCQ by Gruen, Solid State, Liquid Crystal, Quartz, 3 Year Guarantee & Owner's Manual* (New York, n.d.), n.p. in NMAH Collection, catalog #1982.0058.02, Object location, 5012.QU5-9.

3. *Instruction Sheet,* quartz watch, in NMAH collection, catalog #1989.0671.106, Object location 5012.QU8-9.

4. For a discussion of nondeterminism and the digital watch, see Asaf Degani, *Taming HAL: Designing Interfaces Beyond 2001* (New York: Macmillan Palgrave, 2003), 39–47.

5. Chuck Clanton, "The Future of Metaphor in Man-Computer Systems, User Interfaces from Digital Watches to Digital Computers," *BYTE* 8, no. 12 (Dec. 1983), 263–64, 266, 268, 270, 274, 276, 278, 280.

6. *RCA VideoCassette Recorder Owner's Manual, VKP950* (RCA Corporation, 1984), 7, NMAH.

7. "Taming the #!*?@!! VCR: Never in Recording History Have So Few Done So Much for So Many," *People* 37, no. 6 (Feb. 17, 1992), www.people.com/people/archive/article/0,,20112054.00.html, accessed Feb. 2, 2010.

8. Kim Vicente, *The Human Factor: Revolutionizing the Way People Live with Technology* (New York: Routledge, 2006), 191–93.

9. Katie Hafner, "Film Drop-off Sites Fade against Digital Cameras," *New York Times* (Oct. 9, 2007).

10. For "layering" as a key principle in computer programming, see Jerome H. Saltzer and M. Frans Kaashoek, *Principles of Computer System Design: An Introduction* (Burlington, MA: Morgan Kaufmann Publishers, 2009), 24–25.

11. David D. Busch, *David Busch's Nikon D5000 Guide to Digital SLR Photography* (Boston: Course Technology, 2010), 24.

12. Its "shooting modes" included Movie (i.e., video), Portrait, Self-portrait, Landscape, and Night Scene; *Olympus Stylus 300/400 Basic Manual* (n.p., 2003), 24–25.

13. G.W.M., M.D. of New York. *Automobile Dealer and Repairer* 19, no. 2 (Apr. 1915), 51–52.

14. Winton purchaser Charles Henry Russell learned this as early as 1905. "I understand your 1906 car is out," Russell wrote in a letter to the factory in October. "I presume you have made numerous improvements and I am desirous of knowing how I am to get the benefit of those improvements." Winton answered, saying that he could pay for having some of them incorporated into his older car; Henry to Winton, Oct. 10 and 16, 1905, The Staebler Papers, letterpress book, CRH; Staebler

to A. H. Ford, Nov. 1, 1910, about "the proposition which the Reo Motor Car Co. is making to owners of 1910 cars to supply most of the 1911 improvements at a very minimal cost," The Staebler Papers, box 10, letterpress book, Jan. 21, 1910, to May 1911, CRH.

15. For discussion of the economic and intellectual rationale for artificial and planned obsolescence, and of the part played by "consumer engineers" and industrial designers in implementing it in industry, see Jeffrey L. Meikle, *Twentieth Century Limited: Industrial Design in America, 1925–1939* (Philadelphia: Temple University Press, 1979), 70–71.

16. Giles Slade, *Made to Break: Technology and Obsolescence in America* (Cambridge, MA: Harvard University Press, 2006), 43; Slade provides a few examples of intentional product "death dating," as it is sometimes called, the best documented being one involving flashlight bulbs, revealed in a memorandum that came to light in an antitrust case against General Electric, 80–81.

17. Erik Sandberg-Diment, "Should I Buy a Personal Computers [*sic*]?" *New York Times*, Oct. 4, 1983.

18. Slade, *Made to Break*, 30, 261.

19. Sandberg-Diment, "Should I Buy a Personal Computers?"

20. Forrester Research, "The State of Consumers and Technology: Benchmark, 2009," by Charles S. Golvin and Jacqueline Anderson, reported by Jenna Wortham, "The Race to Be an Early Adopter of Technologies Goes Mainstream, a Survey Finds," *New York Times*, Business section, Sept. 2, 2009, B-5.

21. George Booth, "Stand aside, Gruenwald! It's the computer I'm blowing away!" Apr. 29, 1991; Sidney Harris, Elevator choices "Up," "Down," "Other Options," Sept. 9, 1996; Robert Mankoff, "No, I don't want to play chess. I just want you to reheat the lasagna," May 26, 1997, all from *The Complete Cartoons of the New Yorker, Disk Two, 1965–2004* (New York: Advance Magazine Publishers, Inc., 2004).

index

66; Whiting, 75; Willys-Knight, 76; Winton, 57–58, 59, 61, 63, 70, 71, 72, 96, 105, 148
—operation and maintenance of, 8, 54, 62, 88–119, 91, *119G7–G13*, 150, 202; costs of, 85, 86–87, 171; and diagnosis of problems, 8, 54, 111, 122–24, 154–56, 158; and driving safety, 162; and drivers' courses, 92, 98; and driver's posture, 99; and driving techniques, 162–65; for electric cars, 67–68; gasoline for, 110–11, 114, 116, 135–36, 153, 155, 167; gasoline leaks, 129, 130; gasoline vs. other fuels for, 135–36, 251n59; ground clearance of, 61, 70; and hill climbing, 61, 154–55, 163–64; instruction in, 91–93, 98–99; knowledge of, 54, 61, 90–91, 121, 123, 125, 148, 150–51, 152, 161, 167, 169, 189; and lubrication, 131–32, 153, 159; mechanics for, 141; and *Model Garage* articles, 165–70; oil for, 110, 111–13, 116, 132; and oil leaks, 129, 130; and overheating, 111; and radiator leaks, 130–31; and repair and service by owners, 8–9, 75, 126–31; reviews and comparisons of, 170–76; and road hazards, 89, 90–100; seasonal vs. year-round operation of, 115; service facilities for, 9, 121, 202; and servicing by professionals, 141–42, 169; and skidding, 105, 106, 108, 163; and slowing and stopping, 94, 105–9, 154; and sounds, 123; and speed, 8, 61, 164–65; and starting, 93–97, 109, 116, 153–54; and steering, 8, 93, 98–100, 108–9, 155; and steering ratios, 99–100, 108; and supervising performance of, 109–16; and tinkering, 8–9, 120; and tire repair, 127–29; tools for, 124–26, 141, 160; water for, 110, 111, 132; and water leaks, 129–30; winter operation of, 163
—owner's manuals for, 9, 147–62; and audience, 151–62; changing rhetoric of, 150–51; diagnosis of problems with, 124; Ford Model A, 160; Ford Model T, 157–59; General Motors, 157; improved clarity of, 157; learning through, 93; manufacturers' provision of, 9, 77; Oakland, 157; Oldsmobile, 151–56, 157;

organized as catechisms, 148; Studebaker, 159, 160; titling of, 160; and tools, 125; unhelpfulness of, 151–56
Auto Repairing Simplified, 148

bathroom scales, 206
batteries, wet cell, 51, 129, 179. *See also* automobiles: components and functioning of
Bedell, Mary, 144
Best Buy Geek Squad, 202
bicycles, 48–50, 75, 85; adjustment and repair of, 50; and horses, 49; learning to ride, 49–50; makers of, 74; steering of, 98; and tire repair, 50; tools for, 50, 124
boats, 98
books, 149–50. *See also* automobiles: owner's manuals for; personal computers: owner's manuals for; sewing machines: owner's manuals for
Brockmann, John, 40, 156
Buick Owner's Manual, 160

Cadillac Automobile Company, 74
cameras, 51–53, 180, 216; digital, 176, 203, 206, 210–14; film, 210–11; Kodak snapshot, 48, 51–52, 53, 180, 210; movie, 180; Nikon, 211, 212; Olympus, 211, 213–14; Polaroid, 180
cars. *See* automobiles
Carter, Byron, 95
Cellphones. *See* mobile phones
Ceruzzi, Paul, 183
Chalmers (company), 77, 79
Chase, Stuart, 171
Chevrolet (company), 80, 81, 107
Civil War, 30, 34, 147
Claudy, C. H., 69
clocks, 7, 21–29, 30, 48; affordability of, 22; brass movements of, 23; challenges of operating early, 24–29; in colonial America, 21–22; design and fabrication of, 22; development of, 16, 214; digital, 203, 210; and equation of time, 28; industrial production of, 19, 22–23; instructions for, 24–29, 40, 41, 147; keys for winding, 27–28; mass ownership of mechanical, 22; oiling of, 26–27; public, 22; and screws, nuts,

Mom, Gijs, 65, 69
Motor, 60, 148
Motor Car Equipment Company, 125
"Motor More Deadly Than War," 106–7
MP-3 players, 216
"My Best Repair," 126
"My Studebaker Girl," 73

Nash Information for the Owner, 160
National Cash Register Company, 77
National Motor Vehicle Company, 80
Netscape Navigator, 200
New York Times, 176

obsolescence: of digital technologies, 206,
217; and digital vs. analog watches,
208–9; and increased car durability,
142; planned, 214–15; stylistic, 86, 214–
15; technological, 18, 214, 215–17
Oldfield, Barney, 80
Oldsmobile (company), 77, 104
ovens, 178, 206, 210
"Over the Overland Route in an Overland
Car," 73
owner's manuals. *See* automobiles:
owner's manuals for; personal computers: owner's manuals for; personal
technologies: instructions or owner's
manuals for; sewing machines: owner's
manuals for

"Packard and the Ford, The," 72–73
Pagé, Victor W., 102–3, 131, 136–37
PC Brands, 176
peddlers, 14; and automobiles, 74; and
clocks, 19, 23, 26; negative stereotypes
of, 36; and sewing machines, 36–37;
Yankee, 23, 36
personal computers, 1, 20, 181–202;
accessories for, 187; advice about, 5,
181; Apple, 181, 197–98; Apple II, 183,
184, 199; Apple Macintosh, 4, 199,
262n49; applications for, 184; articles,
letters, and product reviews for, 176,
181; and BASIC, 184, 216; and browsers, 200; cartoons about, *119G1*, 217–
18; changes caused by, 185; children's
understanding of, 186; command-
based interface for, 200; command

codes for, 190–91, 193–94; Commodore
PET, 183, 199; and data storage, 6, 183,
194–95; dealers of, 198; development
of, 183; diagnosis of problems of, 189,
197, 201–2; documentation about, 185;
and DOS (disk operating system), 6,
185, 190, 191, 195; evolution of, 199;
first appearance of, 3–4; and floppy
disks, 183, 191, 194–95, 199; frustrations with, 4–7, 177, 181–82; and GUI
(graphical user interface), 190, 199–
200; hard drive of, 194, 199, 200–201,
262n49; IBM, 4, 6, 262n49; IBM-PC,
185, 188, 189–90, 191, 196, 199; keyboard of, 191–92; knowledge of, 6,
183–84, 189, 202; language concerning, 187; learning about, 185, 186–92;
level of ownership of, 184–85, 200, 217;
Microsoft, 190; Microsoft Windows
operating system, 199–200; mouse for,
190, 192, 199, 200; newsletters for, 181;
and operating system, 6, 185, 190, 191,
195, 199–200; operation of, 189–95;
owner's manuals for, 197–98; printers for, 4–6, 7, 187, 188, 193, 196–97;
problems with, 189–91, 200–202; programs for, 6, 183–84, 187, 194, 195; and
pull-down menus, 190; and Quicken,
216; and RAM and ROM, 187; shopping for, 6, 185–88; speed and power
of, 199; starting, 189–90, 194; support
for, 197, 198–99, 202; and syntax errors,
190–91; Tandy TRS-80, 183, 185, 199;
and technical language, 185, 187; and
technological obsolescence, 215–16;
trouble-shooting, 185, 201–2; utility of,
4, 6, 184, 188–89, 200; and VisiCalc,
184, 216; vulnerability of, 194–95; and
word processing, 184, 191–93, 216; and
WordStar, 184, 188, 191, 192, 193, 195,
216; and work vs. leisure, 182
personal technologies: advertising about,
13; challenges of, 3, 7; defined, 2;
demands of, 13, 224–25n10; and digital interface, 206; emergence of, 18;
instructions or owner's manuals for, 7,
13, 19; learning about, 12–13; recent
introduction of, 14; shopping for,
11–12; utility of, 12. *See also* technology

Popular Electronics, 183
Popular Mechanics, 108–9, 162, 165, 173
Popular Science, 162, 165–70, 173
power drills, 178, 179
Predmore, J. Walter, 91
printers. See personal computers: printers for
Prohibition, 142–43
Puritans, 21–22

radios, 117–18, 178, 179, 180–81, 210
Radio Shack, 183, 185
railroads, 22, 28, 57
record players, 209
revenge effect, 3
Richmond, L. L., 36–37
road maps, 114
roads, 57, 61, 73, 107, 108
Russell, Thomas, 111, 112

Sandberg-Diment, Erik, 186, 188, 197, 198, 215, 216, 217
Saturday Evening Post, 9, 93, 162, 163
"Saving the Car by Careful Driving," 105–6
scale, bathroom, 206
scanners, 216
Schlink, F. J., 171
scientific purchasing, 171
Selling Chevrolets, 79, 80, 81
sewing machines, 2, 7, 29–47, 119G2–G6, 214, 217; adjustment of, 43–44; advent of, 19; brands and types of, 33, 35; as commonplace, 48; credit buying of, 30, 32; demonstration of, 34; design and decoration of, 38; door-to-door sales of, 36–37; ease of use of, 34–35; electric vs. treadle, 38; learning to use, 39–42; lubrication of, 45–46; makers of, 44, 230n63, 230n65; manufacturers' claims for, 31–32; marketing and sale of, 37–38; mechanical problems with, 35; needles for, 42; as new, 35–36; operation of, 29, 33, 34–35, 39–40, 43; owner's manuals for, 29, 40–42, 45, 93, 119G3–G6, 147, 156; and paper patterns, 46; parts for, 45, 230n65; public competitions for, 34; reliability of, 45; repairs and maintenance of,

45–46, 119G3–G6; repossession of, 32–33; sale of, 32–33; sales manuals for, 77; and screws, nuts, and bolts, 44; shopping for, 29, 33; Singer, 230n65; "spells" or "fits" of, 35, 121; and technical language, 39; threading of, 42–43; tinkering with, 29; tools for, 43–44, 124; treadle, 39–40
Shiland, Harry, 58
shopping, 10, 179; and eighteenth-century Americans, 14; and technology consumption, 13; and technology vs. traditional consumption, 11–12. See also automobiles; personal computers; personal technologies; sewing machines
Singer, Isaac, 30
Singer Company, 30, 34
Singer Sewing Machine Company, 32, 35–37, 77
Sloan, Alfred E., 215
Society for the Prevention of Cruelty to Apples, 181
spinning wheel, 16
sprinkler systems, 206
Staebler & Sons, 75
Stearns, Myron, 91
Steele, Charles F., 74–75, 76
Stein, Gertrude, 103
Steinbeck, John, 118–19
stoves, 44
Studebaker (company), 104
Study of the Oldsmobile, A, 151–53, 156, 158

"Take Me Out in a Velie Car," 73
taxes, 73
Teape, Vera Marie, 126–27
technical language: and automobiles, 152–53; and clocks, 25–28; and personal computers, 185, 187; and sewing machines, 39
technological determinism, 93
technologies, digital, 203; default settings for, 5, 13, 204–5, 207, 209, 213–14; obsolescence of, 206, 217; range of options of, 206–7
technology: change in, 17–18; consumption of, 10, 11; diffusion of, 47; and dig-